AEROTECH VII

Sustainability in Aerospace Engineering and Technology

IOP Conference Series: Materials Science and Engineering
Volume 405

Putrajaya, Malaysia
7 – 8 August 2018

ISBN: 978-1-5108-7333-9
ISSN: 1757-8981

Printed from e-media with permission by:

Curran Associates, Inc.
57 Morehouse Lane
Red Hook, NY 12571

CURRAN ASSOCIATES INC.
proceedings
.com

Some format issues inherent in the e-media version may also appear in this print version.

Printed by Curran Associates, Inc. (2018)

Institute of Physics
Dirac House, Temple Back
Bristol BS1 6BE UK

Phone: 44 1 17 929 7481
Fax: 44 1 17 920 0979

techtracking@iop.org

Additional copies of this publication are available from:

Curran Associates, Inc.
57 Morehouse Lane
Red Hook, NY 12571 USA
Phone: 845-758-0400
Fax: 845-758-2633
Email: curran@proceedings.com
Web: www.proceedings.com

Table of contents

Volume 405

AEROTECH VII - Sustainability in Aerospace Engineering and Technology

7-8 August 2018, Putrajaya, Malaysia

Accepted papers received: 1 August 2018

Preface

Papers

IOP Conf. Series: Materials Science and Engineering **405** (2018) 011001 doi:10.1088/1757-899X/405/1/011001

AEROTECH VII

Sustainability in Aerospace Engineering and Technology

Peer-reviewed papers from AEROTECH VII Conference,

Putrajaya, Malaysia, August 7-8, 2018.

Main Organizer

Department of Aerospace Engineering, Universiti Putra Malaysia

Co-Organizer

Universiti Kuala Lumpur - Malaysia Institute of Aviation Technology (UniKL MIAT)

IOP Conf. Series: Materials Science and Engineering **405** (2018) 011001 doi:10.1088/1757-899X/405/1/011001

PREFACE

AEROTECH conference was founded in 2005 and first held in Putrajaya, Wilayah Persekutuan. Since then, subsequent AEROTECH conferences were held biennially and this year, it is the seventh edition in the series. AEROTECH conference is organized by Department of Aerospace Engineering, Faculty of Engineering, Universiti Putra Malaysia and it is initiated to provide a platform for both local and international aerospace communities to convene and share the recent development and progress in the various fields of aerospace engineering and technology. For this edition, AEROTECH VII conference is co-organized with Universiti Kuala Lumpur - Malaysia Institute of Aviation Technology. Overall, AEROTECH VII organizer has received about 70 paper submissions from both local and international communities. After extensive review process by at least two experts from the relevant field, 60 papers are accepted to be presented at the conference with 15 of them have been selected for this conference publication.

The aerospace industry has a long and storied history of innovating to meet the societal needs. Today, sustainability and efficiency are the key motivators for aerospace researchers within the industry and academia. Hence, AEROTECH VII adopts the theme "Sustainability in Aerospace Engineering and Technology". The aim of AEROTECH VII is to promote creativity and advancement of the aerospace engineering and technology to meet present and future needs while at the same time, maintaining the spirit of sustainable solutions. Members of academia and industry are expected to exchange ideas and realize new and novel solutions to meet future challenges. AEROTECH VII has attracted researchers and scientists not only from Malaysia but also from countries such as Iraq, Australia, Thailand, Libya, India, Oman and Russia. The conference covers researches in aerospace engineering fields including aerodynamics, propulsion, aerospace structure and material, aerospace design, space systems, flight and system performance, aerospace management and operation, and aerospace reviews.

We, the editors, thank the authors for their contributions and the reviewers for their expertise and also valuable comments. We also like to thank the conference organizing committee and Universiti Putra Malaysia for their commitment and tremendous support to the success of AEROTECH VII. Our deep gratitude is also extended to Universiti Kuala Lumpur - Malaysia Institute of Aviation Technology for their cooperation in co-organizing this year's conference.

Dr. Fairuz Izzuddin Romli
Dr. Mohd Faisal Abdul Hamid
Dr. Chia Chen Ciang

Department of Aerospace Engineering,
Universiti Putra Malaysia, Malaysia.

IOP Conf. Series: Materials Science and Engineering **405** (2018) 011002 doi:10.1088/1757-899X/405/1/011002

Peer review statement

All papers published in this volume of *IOP Conference Series: Materials Science and Engineering* have been peer reviewed through processes administered by the proceedings Editors. Reviews were conducted by expert referees to the professional and scientific standards expected of a proceedings journal published by IOP Publishing.

Remote measurement of cardiopulmonary signal using an unmanned aerial vehicle

Ali Al-Naji[1,2]*, Asanka G. Perera[1] and Javaan Chahl[1,3]

[1] School of Engineering, University of South Australia, Mawson Lakes, Australia
[2] Electrical Engineering Technical College, Middle Technical University, Iraq
[3] Joint and Operations Analysis Division, Defence Science and Technology Group, Melbourne, Australia

*ali_abdulelah_noori.al-naji@mymail.unisa.edu.au

Abstract. This study proposes a computer vision-based system to reveal the cardiopulmonary signal of humans in both static and dynamic scenarios without the need for restricting or contact. The proposed system extracts the signal based on the optical properties of skin colour variations in the facial region using image sequence analysis captured by a hovering drone while being convenient, safe and also cost-effective. The experimental results showed very good agreement, strong correlation coefficients and acceptable error ratios in comparison to reference instruments (finger pulse oximeter and Piezo respiratory belt). Therefore, the proposed system in this paper may be suitable to be applied to detect human physiological parameters in war zones and natural disasters when the contact with the subject is difficult or impossible.

1. Introduction

Measurement of physiological parameters is often essential to maintain individuals' health because of its primary role in the diagnosis of health conditions and monitoring of wellbeing. Current monitoring systems rely on direct contact and require a variety of sensors and electrodes to be attached to patients for long periods, in both clinical and home environments. These systems are restrictive, uncomfortable and potentially unsafe [1-6]. Therefore, there is a need to monitor physiological parameters when contact with skin is either undesirable or is not possible. With rapid growth of remote measurement technology, unobtrusive cardiopulmonary activity monitoring has become feasible.

Interesting research has been undertaken using many remote technologies to detect cardiopulmonary signal, including Doppler effect-based technologies [7-12], thermal imaging technologies [13-18] and imaging photoplethysmography (iPPG) technologies [19-24]. All these remote technologies rely on the observation of physiological and physical variations resulting from the cardiopulmonary activity such as skin colour variations, head motion, arterial pulse motion, thorax motion [25, 26]. However, Doppler and thermal imaging technologies generally need specialised hardware and are somewhat expensive [25, 27]. In addition, Doppler technologies may cause some biological effects on human tissue [13, 25, 27]. In contrast, the iPPG technologies seem to be the most promising methods for remote physiological assessments [25, 27]. But one of the most challenges facing iPPG technologies is when cardiopulmonary signal mix with noise and motion artefacts signal [28]. Many studies proposed different algorithms for noise artefacts removal from the iPPG signal. some examples, including principal component analysis (PCA) [29], independent component analysis (ICA) [22, 23], canonical correlation analysis (CCA.) [19,

30], ensemble empirical mode decomposition (EEMD) [20, 31, 32], chrominance (CHRO) [33], support vector regression (SVR) [21], spatial-subspace rotation (2SR) [34] and partial least squares (PLS) [35]. However, each algorithm has its strengths and weaknesses under different conditions, leading to some challenges and problems that have to be addressed [25].

In this study, a computer vision-based iPPG is proposed to extract the cardiopulmonary signal (heart rate HR and respiratory rate RR) from a hovering unmanned aerial vehicle (UAV) and remove the noise and motion artefacts using a new algorithm using signal decomposition technique based on complete ensemble EMD and ICA technique.

2. Materials and methods

2.1. Research ethics

The research ethics clearance was obtained from the Human Research Ethics Committee (HREC) at the University of South Australia (UniSA). Our research participants had consented to take part in this study and for the use of data that were captured from the UAV after a full explanation of the study procedures. For asset protection and privacy purposes, all collected video data were stored on a secure computer at the UniSA and will not be available for public use.

2.2. Participants and validation methods

This study was performed using 8 participants (4 males and 4 females) aged from 24 to 38 years. The video data were captured using a hovering UAV (3DR solo drone, 10MP, 5.4 mm GoPro Lens) with a resolution of 3840 × 2160 and a frame rate of 25 fps at a distance of 3 meters from the participant as shown in Figure 1. All participants were asked to breathe normally while doing some static and dynamic scenarios (normal talking, blinking, head rotation and facial expression). The validation procedure was carried out considering the measured data provided by a Rossmax SA210 pulse oximeter and MLT1132 Piezo respiratory belt transducer.

Figure 1. Data collection using the UAV (3DR solo drone)

2.3. System framework

Several image and signal processing techniques were considered in the design of the proposed system in this study to extract the relevant features from the facial video signal and eliminate noise artefacts embedded in the image spatial domain. These techniques include skin colour magnification technique, signal decomposition technique, ICA technique, spectral analysis method, band-pass filters and peak detection. The system framework of the proposed system is presented in Figure 2.

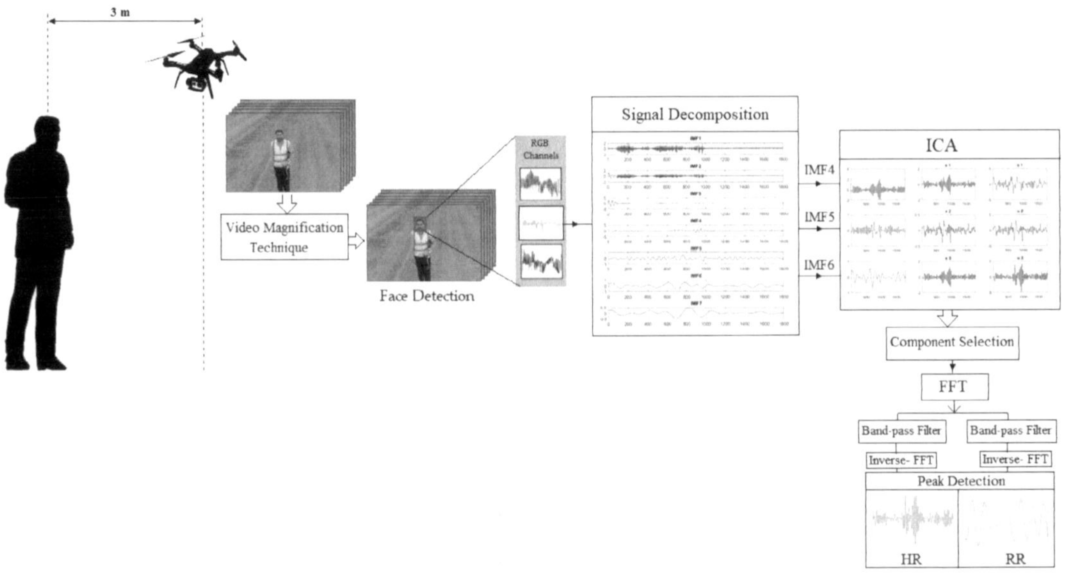

Figure 2. The system framework of the proposed system

2.3.1. Skin colour magnification
The skin colour variations in the participant's face are volumetric changes in the blood caused by the pressure changes during cardiac cycle. The colour of blood varies due to the exchange of gases through the cardiopulmonary activity, which also affects the skin colour. These variations in the skin colour are difficult to recognise. Therefore, a standard video magnification technique [36] was used to magnify these variations. The green (G) component was only selected and magnified 20x with band-pass filtering of 0.15 Hz to 3 Hz since this component contains the most harmonic cardiac signal among the other components (i.e. red and blue) [19, 22, 23, 37].

2.3.2. Face detection
To enhance results accuracy and efficiency, a face detection algorithm proposed by Liao et al. [38], was used in this study to select the facial region of interest (ROI). This face detection algorithm is a robust method to the challenges associated with unconstrained faces (e.g. inclined or angled faces, face rotation and faces in a crowd). The locations of the eyes and mouth for the selected ROI were also removed to reduce the noise artefacts generating from blinking and talking during videoing. The time-series signal, $x\,(t)$, was calculated by averaging of all the image pixel intensities in the selected ROI as follows:

$$x\,(t) = \frac{\sum_{a,b \in ROI} I\,(a,b,t)}{|ROI|} \tag{1}$$

where $I(a,b,t)$ is the pixel intensity at location (a,b) and time (t), and $|ROI|$ is the size of the selected region.

2.3.3. Signal decomposition
Signal decomposition is a powerful representation scheme for multiscale signal and image analysis that is used to get information from the nonstationary time-varying signals. Empirical mode decomposition (EMD) [39] is one of the most commonly used decomposition techniques to reduce the noise artefacts from biomedical signals [40-42]. This technique decomposes a non-linear and/or non-stationary signal into a set of components, namely intrinsic mode functions (IMFs), and selecting the most informative components to represent the original signal. Later on, a noise-assisted signal decomposition technique,

called ensemble EMD (EEMD), was proposed by Wu and Huang [43] to solve the mixing mode problems associated with EMD. However, EEMD has some weaknesses under different assumptions, such as residual noise, reconstruction error and different number of modes at different realisations [44]. Another noise-assisted version of the signal decomposition technique, called complete EEMD, was proposed by Torres et al. [45] to deal with the reconstruction error from the EEMD and to enhance its performance and efficiency by undertaking exact attenuation of the residual noise in the reconstruction process. In this technique, pairs of white noise (positive and negative) are added to the original signal to produce two sets of ensemble IMFs. The procedure for extracting two sets of mixtures, M_1 and M_2, using complete ensemble EMD can be derived by

$$\begin{bmatrix} M_1 \\ M_2 \end{bmatrix} = \begin{bmatrix} 1 & 1 \\ 1 & -1 \end{bmatrix} \begin{bmatrix} x(t) \\ \omega(t) \end{bmatrix} \tag{2}$$

where $x(t)$ is the time-series signal, $\omega(t)$ is the additive while noise, M_1 is the sum of the time-series signal with positive noise and M_2 is the sum of the time-series signal and the negative noise. It is clear from Eq. (2) that the final IMF is the ensemble of both M_1 and M_2 with positive and negative residues of added white noises, thereby reducing residual noise in the reconstructed signal. Figure 3 demonstrates generating seven IMFs of the signal of interest using complete EEMD technique with 100 realisations and 100 iterations.

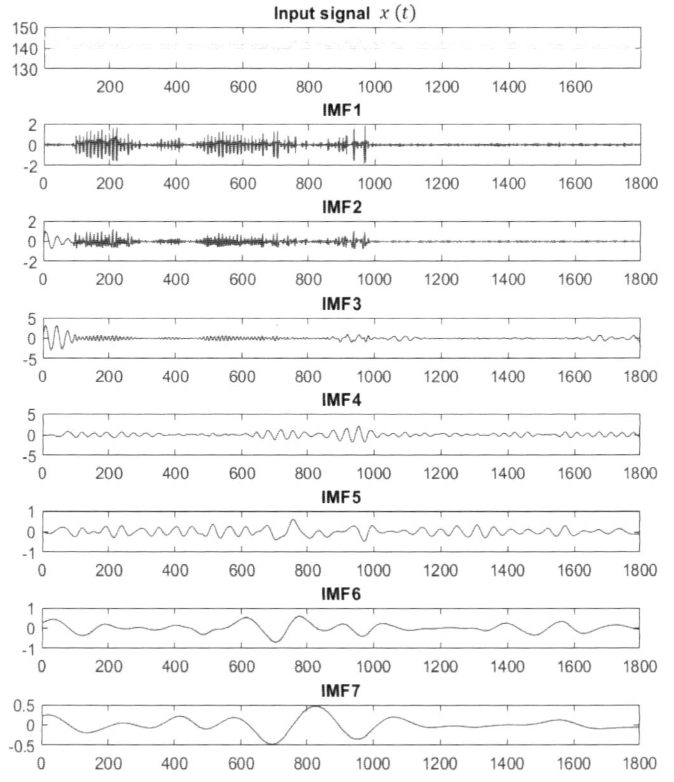

Figure 3. Signal decomposition based on the complete EEMD technique

2.3.4. Independent component analysis (ICA)

ICA is a statistical signal processing technique for solving the blind source separation (BSS) problems and is widely used to remove noise artefacts from biomedical signals [46-48]. ICA-based BSS is used to separate the mixed source signals and retrieve the original sources using high-order statistics. Figure 4 presents the basic scheme diagram of the ICA-based BSS.

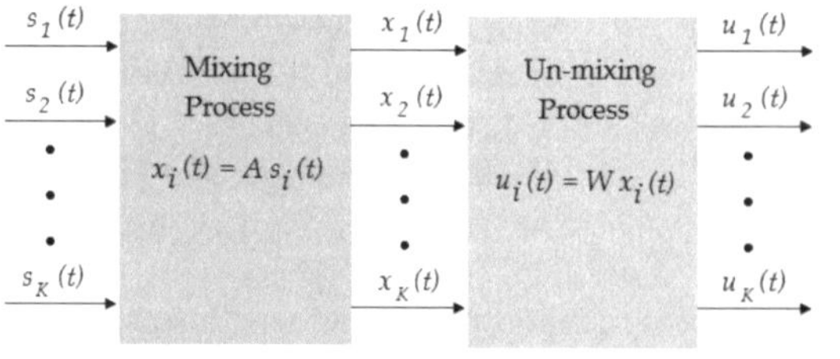

Figure 4. Basic scheme diagram of the ICA-based BSS technique

To understand how ICA operates as the BSS, assume there are K sources independently transmitting signals, $s_i(t) = [s_1(t), s_2(t), \ldots, s_K(t)]$. The observed signals, $x_i(t) = [x_1(t), x_2(t), \ldots, x_K(t)]$, by different sensors, $i = 1,2 \ldots, K$, can be expressed as follows [49]:

$$x_i(t) = As_i(t) = \sum_{i=1}^{K} (a_i s_i(t) + n_i(t)) \quad (3)$$

where A is the unknown mixing matrix with column vectors a_i, and $n_i(t)$ is the additive noise. The observations, $x_i(t)$, are now linear combinations or mixtures of the sources, $s_i(t)$. The ICA can solve the BSS problem under an assumption that all sources are statistically independent of each other. To recover the source signals, $u_i(t)$, the following expression can be used:

$$u_i(t) = Wx_i(t) = W[As_i(t)] \approx s_i(t) \quad (4)$$

where W is the un-mixing weight matrix with column vectors, w_i.

The IMF components with the best cardiopulmonary frequency bands (IMF4, IMF5 and IMF6) were chosen as inputs to ICA technique as demonstrated in Figure 5.

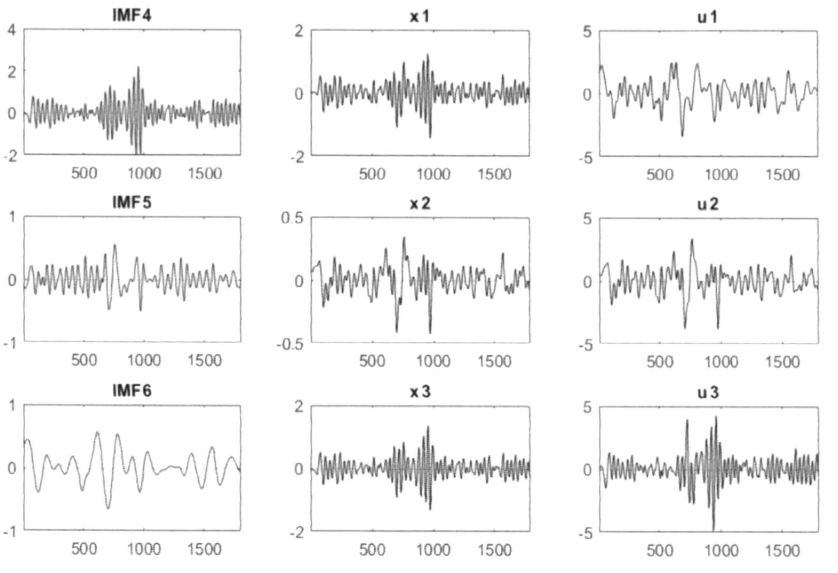

Figure 5. The signals based on ICA technique

The fast Fourier transform (FFT) was used as frequency spectrum analysis to inform the estimation of the cardiopulmonary signal by selecting the best component of the ICA outputs (u3), which has the most resemblance to the expected cardiopulmonary signal. This process was followed by two band-pass filters with selected frequencies of 0.5 to 3 Hz (30-180 beats/min) and 0.15 to 0.4 Hz (9-24 breaths/min). After the filtration, the inverse FFT was applied to obtain the temporal signals of the cardiopulmonary activity. Finally, MATLAB built-in peak detection function [50] was used to determine the number of peaks of the temporal signal and calculated the heart and respiratory rates per minute.

3. Experimental results and discussion

The experimental results and statistical readings of the proposed measurement system for two scenarios (static and dynamic scenarios) were discussed in this section. The limits of agreement (95% confidence intervals), coefficients of correlation and error ratios were calculated using the Bland-Altman plot [51], Pearson's correlation coefficient (PCC), Spearman's rank correlation coefficient (SCC), Kendall rank correlation coefficient (KCC), mean absolute error and root mean square error. All statistical readings were considered when there is no random signal included. The statistics related to HR measurements for the video data captured by the UAV for the static and dynamic scenarios are shown in Figure 6.

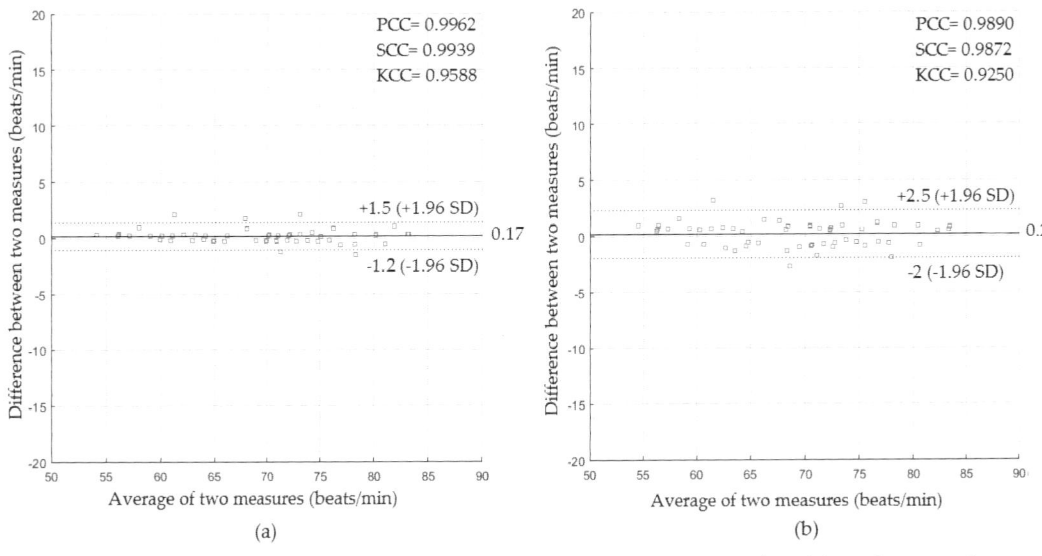

Figure 6. Bland-Altman plot of the cardiac measurements using (a) static scenario, (b) dynamic scenario

Using the data captured from the participants in the static scenario as demonstrated in Figure 6 (a), the mean difference (bias), lower and upper limits of agreement between the cardiac measurements were 0.17, −1.2 and +1.5 beats/min with correlation coefficients of 0.9962, 0.9939, 0.9588 for the PCC, SCC and KCC respectively, and error ratios of 0.52 beats/min and 0.7 beats/min for the mean absolute error and root mean square error, respectively. For the data of the dynamic scenario as demonstrated in Figure 6 (b), the mean difference, lower and upper limit of agreement were 0.22, −2 and +2.5 beats/min with PCC of 0.989, SCC of 0.9872, KCC of 0.925, the mean absolute error of 0.99 beats/min and the root mean square error of 1.16 beats/min.

The statistics related to RR measurements for the video data captured by the UAV for the static and dynamic scenarios are shown in Figure 7.

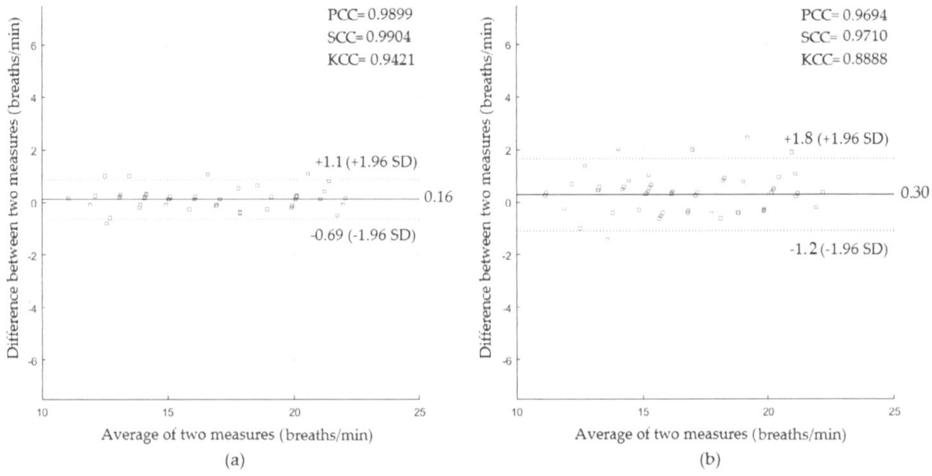

Figure 7. Bland-Altman plot of the respiratory measurements using (a) static scenario, (b) dynamic scenario

Using the data captured from the participants in the static scenario as demonstrated in Figure 7 (a), the mean difference, lower and upper limit of agreement between the respiratory measurements were 0.16, −0.69 and +1.1 breaths/min with correlation coefficients of 0.9899, 0.9904, 0.9421 for the PCC, SCC and KCC respectively, and error ratios of 0.36 breaths/min and 0.46 breaths/min for the mean absolute error and root mean square error, respectively. For the data of the dynamic scenario as demonstrated in Figure 7 (b), the mean difference, lower and upper limit of agreement were 0.3, −1.2 and +1.8 breaths/min with PCC of 0.9694, SCC of 0.971, KCC of 0.8888, the mean absolute error of 0.65 breaths/min and the root mean square error of 0.81 breaths/min.

There are many potential applications of the research presented in this study. Some of the prominent examples include using the system as a lifesaver in disaster zones or warzones, (e.g. field triage of the wounded people), assistance in security and quarantine checkpoints, and also other crowded controlled regions when people are standing or prone with only modest movement. It would also be an important addition to the future robots to be able to detect the physiological parameters of any number of people at different distances.

4. Conclusion

In this study, a computer vision-based system for extracting the human cardiopulmonary signal (HR and RR) from a hovering UAV was proposed. The proposed system used the video magnification technique to magnify the invisible skin colour variations resulting from the pressure changes during the cardiac cycle in the facial region, and applied the complete EEMD technique to decompose the signal of interest into a set of IMF components, while the most informative components were chosen as inputs to the ICA technique and removed the IMF components that fall outside of the frequency bands of interest. The experimental results demonstrated that the proposed system is a feasible solution for remote sensing assessment when the contact with the subject is difficult or impossible in certain situations.

References

[1] Zhao F, Li M, Qian Y, Tsien J Z. Remote measurements of heart and respiration rates for telemedicine. PloS One 2013; 8(10): e71384

[2] De Haan G, Van Leest A. Improved motion robustness of remote-PPG by using the blood volume pulse signature. Physiological Measurement 2014; 35(9): 1913

[3] Butler M, Crowe J, Hayes-Gill B, Rodmell P. Motion limitations of non-contact photoplethysmography due to the optical and topological properties of skin. Physiological Measurement 2016; 37(5): N27

[4] Alnaji A, Gibson K, Chahl J. Remote sensing of physiological signs using a machine vision system. Journal of Medical Engineering and Technology 2017; 41(5): 396-405
[5] Ali A, Chahl J. Contactless cardiac activity detection based on head motion magnification. International Journal of Image and Graphics 2017; 17(1): 1750001
[6] Ali A, Chahl J. Noncontact heart activity measurement system based on video imaging analysis. International Journal of Pattern Recognition and Artificial Intelligence 2017; 31(2): 1-21
[7] Scalise L, Morbiducci U. Non-contact cardiac monitoring from carotid artery using optical vibro-cardiography. Medical Engineering and Physics 2008; 30(4): 490-497
[8] Morbiducci U, Scalise L, De Melis M, Grigioni M. Optical vibrocardiography: a novel tool for the optical monitoring of cardiac activity. Annals of Biomedical Engineering 2007; 35(1): 45-58
[9] Lohman B, Boric-Lubecke O, Lubecke V, Ong P, Sondhi M. A digital signal processor for Doppler radar sensing of vital signs. Engineering in Medicine and Biology Magazine, IEEE 2002; 21(5): 161-164
[10] Kaplan A, OrSullivan J, Sirevaag E J, Lai P, Rohrbaugh J W. Hidden state models for noncontact measurements of the carotid pulse using a laser doppler vibrometer. IEEE Transactions on Biomedical Engineering 2012; 59(3): 744-753
[11] Huang M, Liu J, Xu W, Gu C, Li C, Sarrafzadeh M. A self-calibrating radar sensor system for measuring vital signs. IEEE Transactions on Biomedical Circuits and Systems 2016; 10(2): 352-363
[12] Birsan N, Munteanu D.-P. Non-contact cardiopulmonary monitoring algorithm for a 24 GHz Doppler radar. Annual International Conference of the IEEE in Engineering in Medicine and Biology Society (EMBC), 2012
[13] Abbas A, Heiman K, Orlikowsky T, Leonhardt S. Non-contact respiratory monitoring based on real-time IR-thermography. World Congress on Medical Physics and Biomedical Engineering, 2009
[14] Bennett S L, Goubran R, Knoefel F. Adaptive eulerian video magnification methods to extract heart rate from thermal video. IEEE International Symposium on Medical Measurements and Applications (MeMeA), 2016
[15] Klaessens J H, Born M, Veen A, Sikkens Kraats J, Dungen F A, Verdaasdonk R M. Development of a baby friendly non-contact method for measuring vital signs: First results of clinical measurements in an open incubator at a neonatal intensive care unit. SPIE BiOS, 2014
[16] Murthy R, Pavlidis I, Tsiamyrtzis P. Touchless monitoring of breathing function. 26th Annual International Conference of the IEEE Engineering in Medicine and Biology Society, 2004
[17] Pavlidis I, Dowdall J, Sun N, Puri C, Fei J, Garbey M. Interacting with human physiology. Computer Vision and Image Understanding 2007; 108(1): 150-170
[18] Sun N, Garbey M, Merla A, Pavlidis I. Imaging the cardiovascular pulse. IEEE Computer Society Conference in Computer Vision and Pattern Recognition, 2005
[19] Alnaji A, Perera A G, Chahl J. Remote monitoring of cardiorespiratory signals from a hovering unmanned aerial vehicle. BioMedical Engineering OnLine 2017; 16(1): 101
[20] Chen D, Wang J, Lin K, Chang H, Wu H, Chen Y, Lee S. Image sensor-based heart rate evaluation from face reflectance using Hilbert–Huang Transform. IEEE Sensors Journal 2015; 15(1): 618-627
[21] Hsu Y, Lin Y, Hsu W. Learning-based heart rate detection from remote photoplethysmography features. IEEE International Conference on Acoustics, Speech and Signal Processing, 2014
[22] Poh M, McDuff D, Picard R W. Non-contact, automated cardiac pulse measurements using video imaging and blind source separation. Optical Society of America 2010; 18(10): 10762-10774
[23] Poh M, McDuff D, Picard R W. Advancements in noncontact, multiparameter physiological measurements using a webcam. IEEE Transactions on Biomedical Engineering 2011; 58(1): 7-11
[24] Qi H, Guo Z, Chen X, Shen Z, Wang Z. Video-based human heart rate measurement using joint

blind source separation. Biomedical Signal Processing and Control 2017; 31: 309-320
[25] Alnaji A, Gibson K, Lee S, Chahl J. Monitoring of cardiorespiratory signal: Principles of remote measurements and review of methods. IEEE Access 2017; 5: 15776-15790
[26] Alnaji A, Chahl J. Detection of cardiopulmonary activity and related abnormal events using Microsoft Kinect sensor. Sensors 2018; 18(3): 920
[27] Kranjec J, Beguš S, Geršak G, Drnovšek J. Non-contact heart rate and heart rate variability measurements: A review. Biomedical Signal Processing and Control 2014; 13: 102-112
[28] Alnaji A, Chahl J. Remote optical cardiopulmonary signal extraction with noise artifact removal, multiple subject detection and long-distance. IEEE Access 2018; 6: 11573 - 11595
[29] Lewandowska M, Rumiński J, Kocejko T, Nowak J. Measuring pulse rate with a webcam—a non-contact method for evaluating cardiac activity. Federated Conference on Computer Science and Information Systems (FedCSIS), 2011
[30] Alnaji A, Chahl J. Simultaneous tracking of cardiorespiratory signals for multiple persons using a machine vision system with noise artifact removal. IEEE Journal of Translational Engineering in Health and Medicine 2017; 5: 1-10
[31] Lin K, Chen D, Tsai W. Face-based heart rate signal decomposition and evaluation using multiple linear regression. IEEE Sensors Journal 2016; 16(5): 1351-1360
[32] Cheng J, Chen X, Xu L, Wang Z J. Illumination variation-resistant video-based heart rate measurement using joint blind source separation and ensemble empirical mode decomposition. IEEE Journal of Biomedical and Health Informatics 2017; 21(5): 1422-1433
[33] De Haan G, Jeanne V. Robust pulse rate from chrominance-based rPPG. IEEE Transactions on Biomedical Engineering 2013; 60(10): 2878-2886
[34] Wang W, Stuijk S, De Haan G. A novel algorithm for remote photoplethysmography: spatial subspace rotation. IEEE Transactions on Biomedical Engineering 2016; 63(9): 1974-1984
[35] Xu L, Cheng J, Chen X. Illumination variation interference suppression in remote PPG using PLS and MEMD. Electronics Letters 2017; 53(4): 216-218
[36] Wu H, Rubinstein M, Shih E, Guttag J, Durand F, Freeman W. Eulerian video magnification for revealing subtle changes in the world. ACM Trans. Graph. 2012; 31(4): 65
[37] Verkruysse W, Svaasand L O, Nelson J S. Remote plethysmographic imaging using ambient light. Optics Express 2008; 16(26): 21434-21445
[38] Liao S, Jain A K, Li S Z. A fast and accurate unconstrained face detector. IEEE Transactions on Pattern Analysis and Machine Intelligence 2016; 38(2): 211-223
[39] Huang N E, Shen Z, Long S R, Wu M C, Shih H H, Zheng Q, Yen N, Tung C C, Liu H H. The empirical mode decomposition and the Hilbert spectrum for nonlinear and non-stationary time series analysis. Proceedings of the Royal Society of London A: Mathematical, Physical and Engineering Sciences, 1998
[40] Blanco-Velasco M, Weng B, Barner K E. ECG signal denoising and baseline wander correction based on the empirical mode decomposition. Computers in Biology and Medicine 2008; 38(1): 1-13
[41] Chen X, Liu A, Chiang J, Wang Z J, McKeown M J, Ward R K. Removing muscle artifacts from EEG data: Multi-channel or single-channel techniques? IEEE Sensors Journal 2016; 16(7): 1986-1997
[42] Zhang X, Zhou P. Filtering of surface EMG using ensemble empirical mode decomposition. Medical Engineering and Physics 2013; 35(4): 537-542
[43] Wu Z, Huang N E. Ensemble empirical mode decomposition: a noise-assisted data analysis method. Advances in Adaptive Data Analysis 2009; 1(1): 1-41
[44] Colominas M, Schlotthauer G, Torres M E. Improved complete ensemble EMD: A suitable tool for biomedical signal processing. Biomedical Signal Processing and Control 2014; 14: 19-29
[45] Torres M E, Colominas M A, Schlotthauer G, Flandrin P. A complete ensemble empirical mode decomposition with adaptive noise. IEEE International Conference on Acoustics, Speech and Signal Processing (ICASSP), 2011

[46] Safieddine D, Kachenoura A, Albera L, Birot G, Karfoul A, Pasnicu A, Biraben A, Wendling F, Senhadji L, Merlet I. Removal of muscle artifact from EEG data: comparison between stochastic (ICA and CCA) and deterministic (EMD and wavelet-based) approaches. EURASIP Journal on Advances in Signal Processing 2012; 2012(1): 127
[47] Srivastava G, Crottaz-Herbette S, Lau K, Glover G, Menon V. ICA-based procedures for removing ballistocardiogram artifacts from EEG data acquired in the MRI scanner. Neuroimage 2005; 24(1): 50-60
[48] Zhou W, Gotman J. Removal of EMG and ECG artifacts from EEG based on wavelet transform and ICA. 26th Annual International Conference of the IEEE Engineering in Medicine and Biology Society, 2004
[49] Oja E. Applications of independent component analysis. Neural Information Processing, 2004
[50] Billauer E. Peak detection using MATLAB. Dostopno na: http://billauer.co.il/peakdet.html, 2011
[51] Bland J M, Altman D G. Statistical methods for assessing agreement between two methods of clinical measurement. International Journal of Nursing Studies 2010; 47(8): 931-936

IOP Conf. Series: Materials Science and Engineering **405** (2018) 012002 doi:10.1088/1757-899X/405/1/012002

Computational study of low and high subsonic speed aerodynamic characteristics of the modified airfoil profile

Tosaporn Soontornpasatch

Department of Mechanical and Aerospace Engineering, Faculty of Engineering, King Mongkut's University of Technology North Bangkok, Thailand.

tosaporns@kmutnb.ac.th

Abstract. The computational study is conducted to evaluate the low and high subsonic speeds aerodynamic characteristics of the modified airfoil profile. This profile is created using the mean-camber line of the Eppler-197 airfoil profile and the NASA LS-0013 symmetrical profile. The ANSYS FLUENT, a commercial computational fluid dynamic code, is applied to calculate the aerodynamic parameters. At low speed condition, the study is conducted at Reynolds number from 1.5×10^6 to 6.0×10^6 and angle of attack is varied from -4° to 18°. At high speed condition, the Mach number is varied from 0.4 to 0.7 while the Reynolds number is fixed at 9.0×10^6. The range of angle of attack is taken as -2° to 6°. The results indicate that the modified profile has better characteristics than the original profile. At low speed range, the maximum lift coefficient and maximum lift-to-drag ratio are higher than the original profile. However, the minimum drag coefficient is lower. The characteristics of the modified profile at high subsonic speed is also better than the original profile. At Mach number more than 0.64, the lift coefficient and lift-to-drag ratio are also higher while drag coefficient is lower when compared with the original profile.

1. Introduction

This study is a continuous work from Ref. [1], which has studied the aerodynamic characteristics of modified airfoil profile, Eppler-197 at low subsonic speed. In the study, this profile has been modified by using some symmetrical profiles. One of these profiles is NASA LS (1)-0013. The result from the study has shown that the modified profile with NASA LS (1)-0013 profile has a pressure coefficient distribution as the same pattern as NASA supercritical airfoil (which is designed for the high subsonic speed). Therefore, this present study is set up to evaluate high subsonic aerodynamic characteristics of this modified profile. Moreover, the previous study has been done at Reynolds number 1×10^6 only. Then, this work has been also extended to study at other Reynolds numbers at low speed range. The modification airfoil profile for high speed applications is not a new concept. There are many airfoil profiles that are modified in order to improve their characteristics in high subsonic range [2-4]. In the past, these studies have been performed using wind tunnel as a primary tool. However, in the modern day, this task can be performed by numerical method or CFD code instead. There are many research works that have used CFD code to study the aerodynamic parameters at high subsonic range [5,6]. Therefore, the commercial CFD code, ANSYS FLUENT, is used as a primary tool in this study.

The aim of this study is to evaluate the aerodynamic characteristics of the modified Eppler-197 airfoil profile at the range of low and high speeds. The lift coefficient, drag coefficient and lift-to-drag ratio are the primary variables in this study. At low speed condition, Reynolds number is varied from 1.5×10^6 to

IOP Conf. Series: Materials Science and Engineering **405** (2018) 012002 doi:10.1088/1757-899X/405/1/012002

6.0×10^6 (based on the chord length) and the angle of attack varied from -4° to 18°. At high speed condition, the Reynolds number is fixed at 9.0×10^6 and the Mach number is varied from 0.4 to 0.7. The angle of attack is varied from -2° to 6°. The data of Eppler-197 original profile also comes from the simulation in this study too.

2. Airfoil modification

The mean-camber line of the Eppler-197, a low speed airfoil profile, is used as a basic camber line for the modified profile. Coordinate of this mean-camber line can be calculated by using Eqn. 1, where y is the coordinate of the Eppler-197 profile and subscript u and l mean the upper and lower surface, respectively.

$$y_C = \frac{y_u + y_l}{2} \tag{1}$$

The coordinate of the symmetrical NASA LS-0013 airfoil profile [7] is used to create the surface around the mean-camber line of the modified profile. This surface is calculated using the same method in Ref. [1]. The comparison between the original and modified profile is shown in Figure 1.

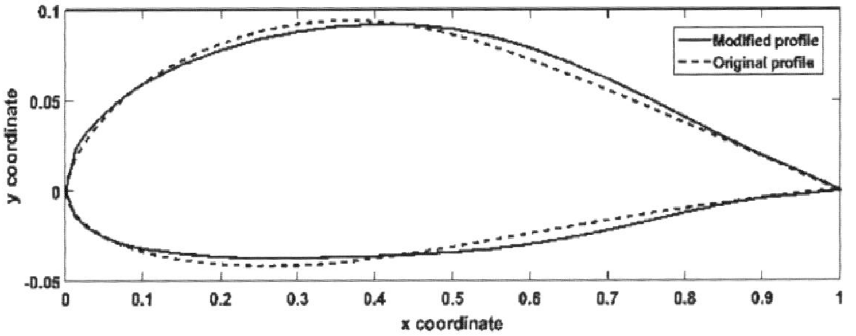

Figure 1. Comparison between the original and modified profile

3. Numerical model

The control volume around the airfoil is the C-H type domain as depicted in Figure 2. The horizontal and vertical length of the domain is 120 times and 100 times of the airfoil chord length, respectively. The internal area of the domain is divided by using structural mesh scheme. Mesh sensitivity testing is used to determine the number of meshes in the domain.

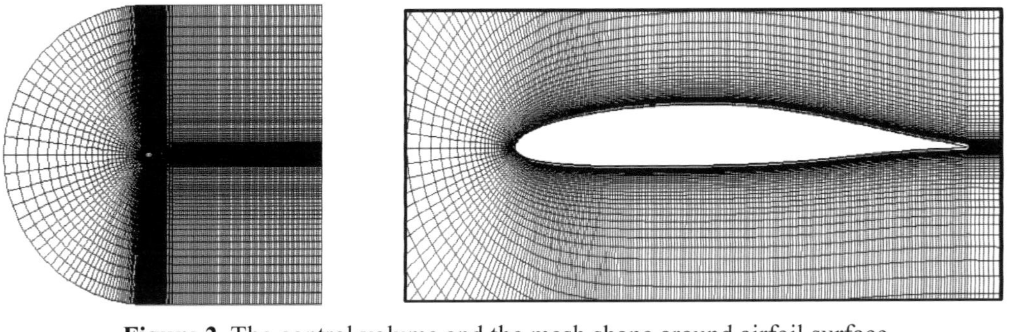

Figure 2. The control volume and the mesh shape around airfoil surface

The computational prediction is performed using commercial computational fluid dynamic code, ANSYS FLUENT 17.2. This program performs the calculation by solving the governing equations (continuity, momentum and energy) and the equation of state using the finite volume method. In order to deal with the compressibility effect in the flow field, the density-based solver (DBS) is used in this

IOP Conf. Series: Materials Science and Engineering **405** (2018) 012002 doi:10.1088/1757-899X/405/1/012002

study and the second order upwind scheme is used to calculate flow parameters. The program is set to perform the calculation until all residual values of the flow variables are lower than 4×10^{-4} and all of the aerodynamic parameters are unchanged.

In the control volume, air is selected as a medium and is assumed to be an ideal gas. The pressure farfield boundary condition is applied at outer edge of the domain. The temperature (T) is set at 300 K. In order to keep the Reynolds number at 9.0×10^{6} while Mach number is varied, the density (ρ_∞) has to be changed at each Mach number. The density value is calculated using Eqn. 2 and the static pressure (P) value, which is applied at the domain boundary can be calculated by Eqn. 3. The Re is Reynolds number, μ_∞ is dynamics viscosity (kg.m/s), U_∞ is freestream velocity (m/s), c is chord length (m) that is fixed at 1 m in this study, and R is 0.287 (kJ/kg.K).

$$\rho_\infty = \frac{Re\,\mu_\infty}{U_\infty c} \tag{2}$$

$$P = \rho_\infty RT \tag{3}$$

At the surface of the airfoil, it is assumed that there is no relative velocity between the surface and air. Therefore, no-slip boundary is applied on the airfoil surface. The roughness of the surface is set at 0 m. The $k-\omega$ SST turbulence model is applied in this study because this model has more accuracy, higher numerical stability and can be used very well in compressible flow condition and when adverse pressure gradient occurs in the flow field [8]. In order to capture the viscous sub-layer in boundary, the y+ value should be less than 5 [10,11]. The y+ distance wall distance is estimated using Eqn. 4 [12]. From the estimation, the wall distance is 1.624×10^{-5} m.

$$C_f = 0.058\,(Re_X)^{-0.2} \tag{4}$$

4. Mesh sensitivity and validation

The number of meshes is determined by mesh sensitivity testing. The experiment data of NACA0012-64 airfoil profile [9] is used for the test. The calculation is performed at Mach number 0.4 and angle of attack is 2°. The number of meshes is varied from 5,482 to 203,963. The results indicate that the lift coefficient and drag coefficient do not change when the number of meshes is more than 127,846 and 94,836 as being shown in Figure 3 and Figure 4, respectively. At this point, the value of lift and drag coefficient from the calculation are 0.193 and 0.0143, which agree well with the experiment data in Ref. [9].

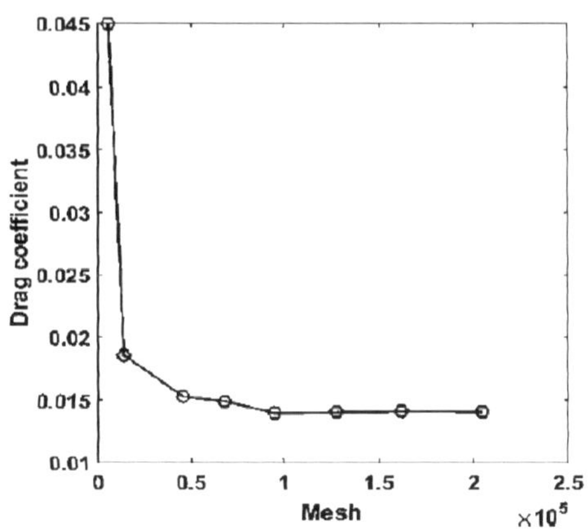

Figure 3. The relation between drag coefficient and the number of meshes

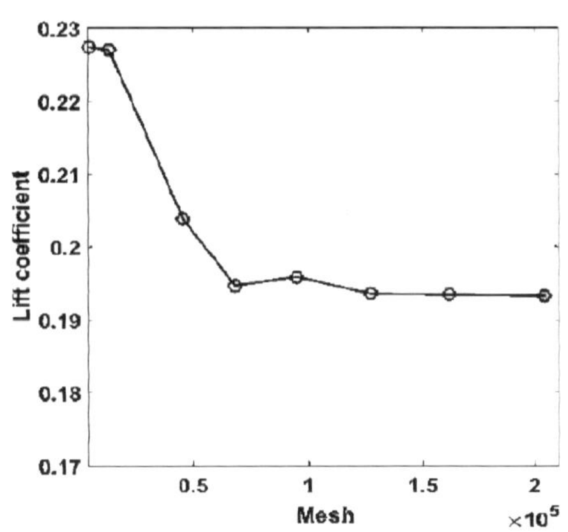

Figure 4. The relation between lift coefficient and the number of meshes

IOP Conf. Series: Materials Science and Engineering **405** (2018) 012002 doi:10.1088/1757-899X/405/1/012002

The test is further evaluated by varying the Mach number from 0.4 to 0.7 at the same angle of attack. The result shown in Figure 5 agrees very well with the experiment data. At the 127,846 meshes, the y+ value along the upper and bottom surface of an airfoil varies from 0.010 to 0.065 as shown in Figure 6. Therefore, the condition that y+ should be less than 5 is satisfied and the number of meshes 127,846 is used in this study.

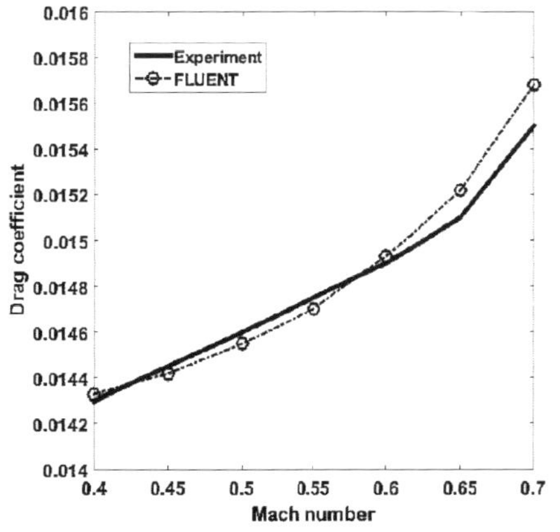

Figure 5. Comparison of drag coefficient value between experiment data and simulation

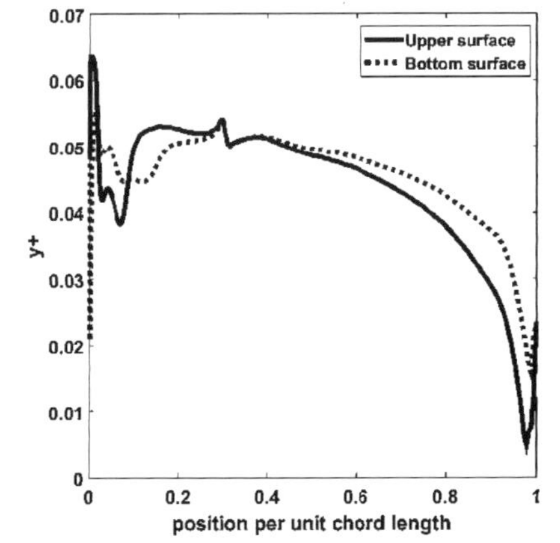

Figure 6. The y+ value along chord length of NACA0012-64

5. Results and discussion

5.1. Low speed condition

The relation between angle of attack and lift coefficient of the modified profile shows in Figure 7. The zero lift angle of attack is set at −3.68° for all Reynolds numbers, which is nearly the same value in Ref. [1]. The slope of lift coefficient curve increases with increasing Reynolds number. The value of drag coefficient tends to decrease when Reynolds number increases as shown in Figure 8.

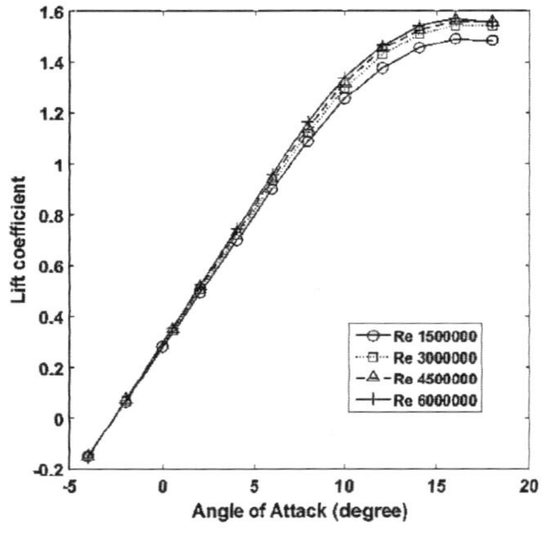

Figure 7. The relation between lift coefficient and angle of attack

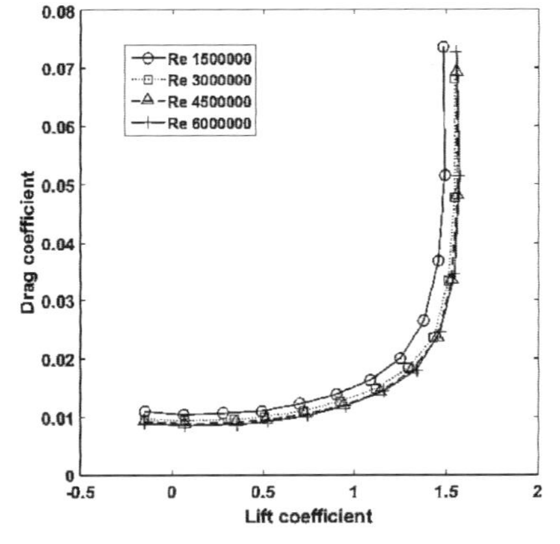

Figure 8. The relation between drag coefficient and lift coefficient

Lift-to-drag ratio increases rapidly from Reynolds number 1.5×10^6 to 3.0×10^6 as shown in Figure 9. However, this value does not change much at Reynolds numbers from 4.5×10^6 to 6.0×10^6. When compared with the original profile, the minimum drag coefficient of the modified profile is lower at all Reynolds number range. However, the maximum lift coefficient of the modified profile is higher than

IOP Conf. Series: Materials Science and Engineering **405** (2018) 012002 doi:10.1088/1757-899X/405/1/012002

the original profile at all Reynolds number. The maximum lift-to-drag ratio of the modified profile is higher than the original profile at Reynolds number between 3×10^6 and 6×10^6. The relations are shown in Figure 10 to Figure 12.

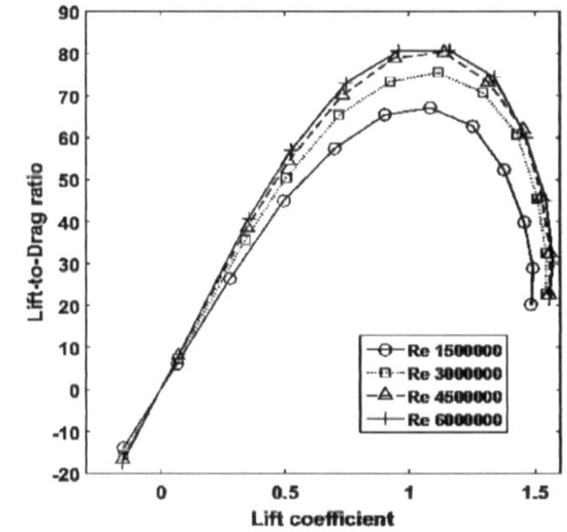

Figure 9. The relation between lift-to-drag ratio and lift coefficient

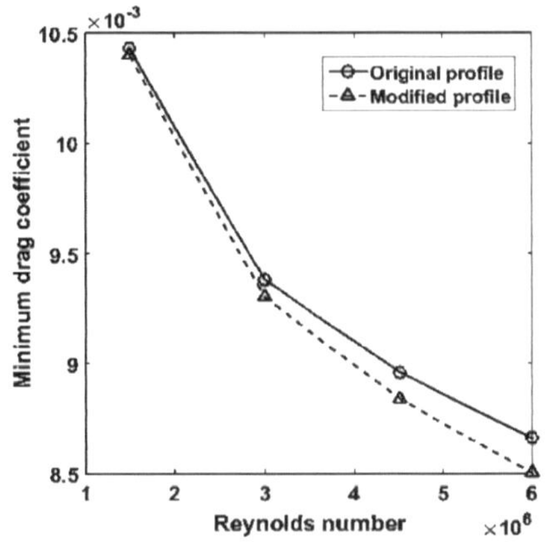

Figure 10. Minimum drag coefficient at each Reynolds number

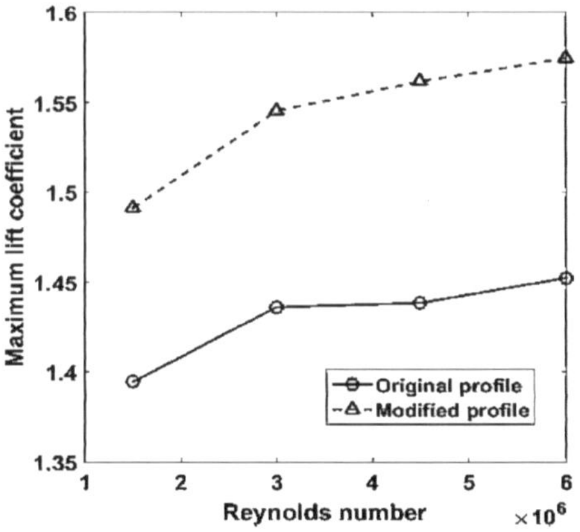

Figure 11. Maximum lift coefficient at each Reynolds number

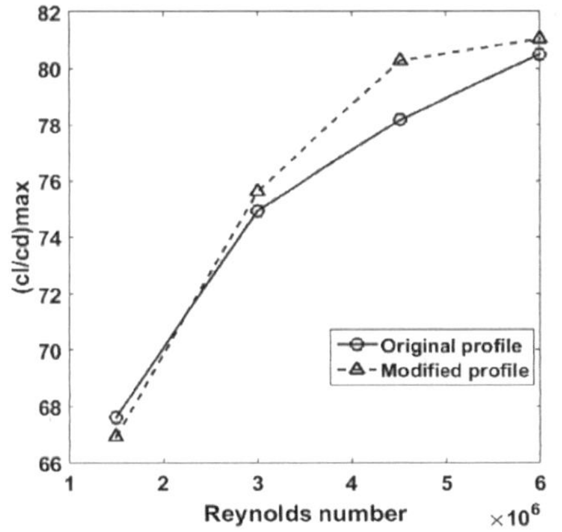

Figure 12. Maximum lift-to-drag ratio at each Reynolds number

5.2. High speed condition

The relation between lift coefficient and angle of attack at Mach number from 0.4 to 0.7 is shown in Figure 13. At Mach number lower than 0.64, lift coefficient linearly increases when angle of attack increases and the lift coefficient value of the modified profile is nearly the same as the original profile. However, at Mach number 0.64 and 0.7, the lift coefficient increases in non-linear pattern for the angle of attack above 4° and 0°, respectively. At these two points, the lift coefficient of the modified profile increases higher than the value from the original profile. The relation between lift and drag coefficient shows in Figure 14. At Mach number between 0.4 to 0.6, drag coefficient of the modified and original profile are at the same level at all lift coefficient range. When Mach number is more than 0.64, drag coefficient of the modified profile is lower than the original profile and the increase rate is also lower.

The relation between lift coefficient and lift-to-drag ratio shows in Figure 15. It shows that the value of lift-to-drag ratio tends to decrease when Mach number increases. At Mach number lower than 0.64, the value of lift-to-drag ratio of the modified profile and original profile are nearly the same at all lift coefficient range. However, at Mach number more than 0.64, the lift-to-drag ratio of the modified profile is higher than the original profile. Figure 16 shows the comparison of pressure coefficient distribution between original and modified profile at Mach number 0.74 and angle of attack 4°. For the original profile, it shows that the normal shock wave occurs at 0.38 of the chord length on the upper surface. However, the shock wave of the modified profile occurs at 0.43 of the chord length, which is far from the leading edge more than the original profile. The delay of the normal shock wave on the modified profile causes the smaller area of separation flow behind the shock wave, as shown in Figure 17, and causes higher lift coefficient and lower drag coefficient at high Mach number as previously shown in Figure 13 and Figure 14.

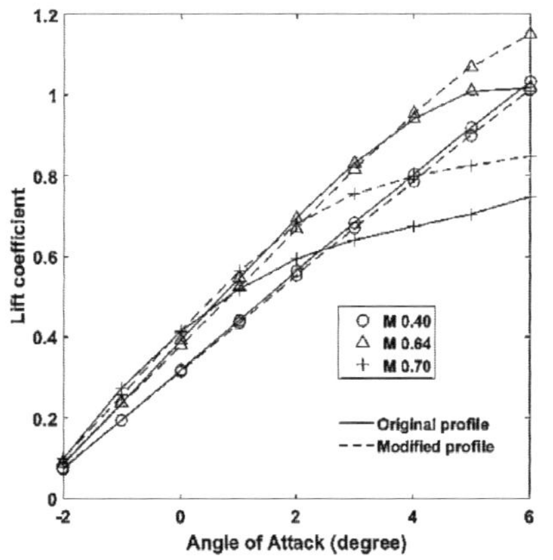

Figure 13. Lift coefficient at each angle of attack at various Mach numbers

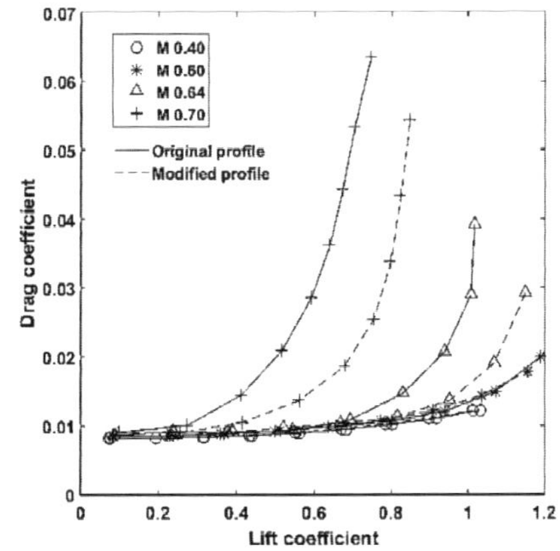

Figure 14. Relation between drag coefficient and lift coefficient at various Mach numbers

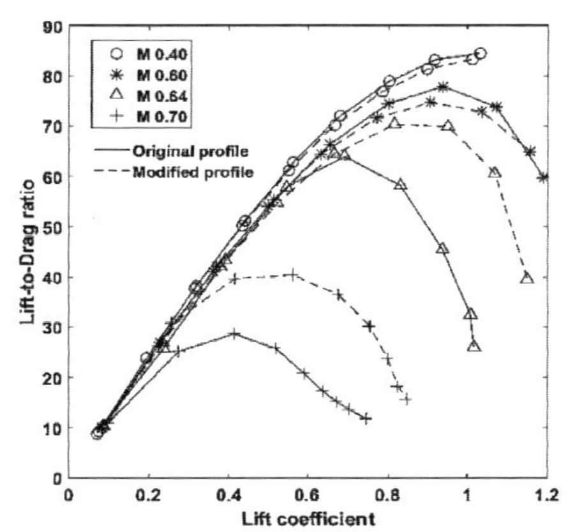

Figure 15. Relation between lift-to-drag ratio and lift coefficient at various Mach numbers

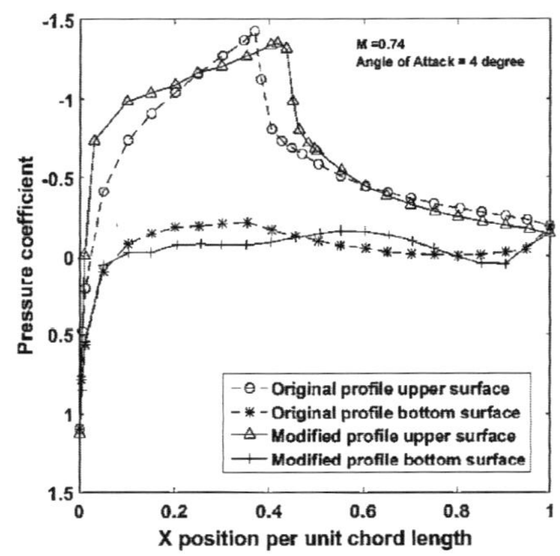

Figure 16. Comparison of the pressure distribution between original and modified profile at M = 0.74 and angle of attack of 4°

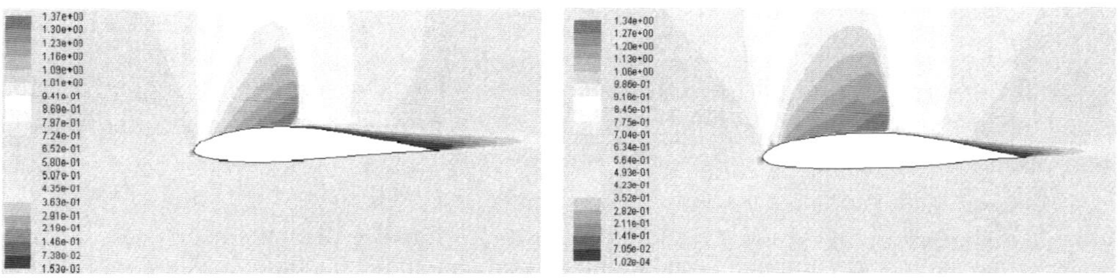

Figure 17. Comparison of the Mach contour between original (left) and modified (right) profile

6. Conclusion

The obtained results of this study lead to the conclusion that the modified Eppler-197 profile has better aerodynamic characteristics when compared with the original profile. At low speed range, maximum lift coefficient and maximum lift-to-drag ratio of the modified profile are higher than those for the original profile. However, the minimum drag coefficient is lower. The effective range of the modified profile is at Reynolds number between 3.0×10^6 and 6.0×10^6, which all aerodynamic parameters are better than the original profile. At high speed, the modified profile also has better characteristics than the original profile when Mach number is higher than 0.64. The lift coefficient and lift-to-drag ratio have higher value while the drag coefficient is lower when compared with the original profile.

References

[1] Soontornpasatch T. Analysis of flow field and characteristics of the modified Eppler-197 airfoil profile. 2nd International Conference on Engineering Science and Inovative Technology (ESIT) 2016, Phuket, Thailand
[2] Laurence K, Loftin Jr. Aerodynamic characteristics of NACA 64-010 and 0010-1.10 40/1.051 airfoil sections at Mach number from 0.3 to 0.85 and Reynolds number from 4×10^6 to 8×10^6. NACA TN No. 3244, 1954
[3] Polentz P P. Comparison of the aerodynamic characteristics of the NACA0010 and 0010-64 airfoil sections at high subsonic Mach number. NACA RM No. A9G19, 1949
[4] Hemenover A D. The effects of camber on the variation with Mach number of the aerodynamic characteristics or a 10-percent-thick modified NACA four-digit-series airfoil section. NACA TN No. 2998, 1953
[5] Holst T L. Computational fluid dynamics drag prediction-results from the Viscous Transonic Airfoil workshop. NASA TM No.100095, 1988
[6] Mani M, Rider B, Sclafani A, Winkler C, Vassberg J, Dorgan A J, Cary A, Tinoco E. Reynolds-averaged Navier-Stokes technology for transonic drag prediction: A Boeing perspective. Journal of Aircraft 2014; 51(4): 1118-1134
[7] Ferris J C, McGhee R J, Barnwell R W. Low-speed wind-tunnel results for symmetrical NASA LS(1)-0013 airfoil. NASA TM No. 4003, 1987
[8] Blazek J. Computational Fluid Dynamics: Principle and Applications. Elsevier, 2005
[9] Ferri A. Complete turbulation in the United States of the test of 24 airfoils at high Mach number. NACA ACR NO. L5E21, 1945
[10] Tennekes H, Lumley J L. A first course in turbulence 15th printing. The MIT Press, 1994
[11] Ariff M, Salim S M, Cheah S C. Wall Y+ approach for dealing with turbulent flow over a surface mounted cube: Part 1 low Reynolds number. International Conference on CFD in the Minreals and Process Industries CSIRO, Melbourne, Australia, 2009
[12] Schlichting H. Boundary-Layer Theory. McGraw-Hill, 1979

IOP Conf. Series: Materials Science and Engineering **405** (2018) 012003 doi:10.1088/1757-899X/405/1/012003

Human Detection and Motion Analysis from a Quadrotor UAV

Asanka G Perera[1], Ali Al-Naji[1,2], Yee Wei Law[1], and Javaan Chahl[1,3]

[1] School of Engineering, University of South Australia, Mawson Lakes, SA 5095, Australia.
[2] Electrical Engineering Technical College, Middle Technical University, Al Doura 10022, Baghdad, Iraq.
[3] Joint and Operations Analysis Division, Defence Science and Technology Group, Melbourne, Victoria 3207, Australia.

E-mail: asanka.perera@mymail.unisa.edu.au

Abstract. This work focuses on detecting humans and estimating their pose and trajectory from an umnanned aerial vehicle (UAV). In our framework, a human detection model is trained using a Region-based Convolutional Neural Network (R-CNN). Each video frame is corrected for perspective using projective transformation. Using Histogram Oriented Gradients (HOG) of the silhouettes as features, the detected human figures are then classified for their pose. A dynamic classifier is developed to estimate forward walking and a turning gait sequence. The estimated poses are used to estimate the shape of the trajectory traversed by the human subject. An average precision of 98% has been achieved for the detector. Experiments conducted on aerial videos confirm our solution can achieve accurate pose and trajectory estimation for different kinds of perspective-distorted videos. For example, for a video recorded at 40m above ground, the perspective correction improves accuracy by 37.1% and 17.8% in pose and viewpoint estimation respectively.

1. Introduction

Unmanned aerial vehicles (UAVs) can be deployed in a variety of search and rescue, and surveillance applications by leveraging its mobility and operational simplicity. In some situations, a UAV's ability to recognize the actions of a human subject is desirable, then take responsive actions. Recognizing human actions from videos captured from a static platform is a challenging task owing to the articulated structure and range of possible poses of the human body. Recognition is further challenged by the quality of videos which include perspective distortion, occlusion, and motion blur.

The study presented in this paper is focused on using a UAV to recognize human subjects from an aerial video and to estimate the gait sequence and movement trajectory. Our solution consists of the following steps: (i) The human detector is trained using the method of Region-based Convolutional Neural Network (R-CNN) [1] with aerial images selected from publicly available aerial image datasets and our field images. (ii) The perspective correction step compensates for perspective distortion in aerial images.

Multiple pre-annotated homography matrices are used for different levels of distortion caused by different camera elevation angles. The experimental results show that this technique enhances performance in gait and trajectory estimation for aerial videos. (iii) The segmentation step generates the silhouettes and uses Histograms of Oriented Gradients (HOG) [2] as feature descriptors. (iv) The pose estimation uses a dynamic classifier inspired by [3, 4]. (v) The trajectory estimation step estimates the shape of the human subject's path using 3-D skeletons and localizing them with respect to the initial pose and viewpoint.

The key contribution of this paper is a preliminary solution that a vision-capable quadrotor will be able to use for human detection, pose estimation and trajectory estimation. This study proposes to use an R-CNN detector and a perspective correction module in combination with a novel dynamic classifier architecture. Unlike other designs, our classifier uses temporal relationships between poses to achieve efficient pose and trajectory estimations.

2. Related work

Aerial videos are always subject to some level of perspective distortion due to their aerial viewpoint. It is necessary for videos to be perspective-corrected before classification. Projective transformation, or homography, is a standard technique for correcting perspective distortion [5], but this traditional approach requires the *vanishing point* to be manually specified. Rogez et al. [6] used manually determined *vertical scene lines* to estimate the vanishing point and localize the reconstructed poses based on the vanishing point. Our homography step is similar to theirs, the difference being that we determine the vanishing point based on the altitude and angle of the camera.

Projective transformation has been used to achieve improved results for videos from overhead cameras [7, 8]. In [7], affine transformation has been applied using the 3D scene information for perspective correction. Their experiments show that perspective correction has a noticeable impact on recognition performance. Li et al. [8] reported a human detection accuracy of 87.2% for CAVIAR dataset [9] when the images are corrected for perspective using 3D scene information. When using the 2D search and in-plane rotation the accuracy was only 38.3%. This 48.9% improvement was achieved for the static videos while our 37.1% and 17.8% accuracy improvements in pose and viewpoint estimation were achieved from dynamic videos (40m above ground).

Dynamic classifier selection (DCS) [10] is based on the *local accuracy estimation* of each individual classifier. The main idea is selecting an individual classifier which is most likely correct for a given sample. The final classification decision is made only by the selected classifier. A relatively similar classifier was developed by Ko et al. [4] by integrating a majority voting system. Our classifier follows the DCS principles, but it selects the best individual classifier without executing the entire ensemble of classifiers.

Using UAVs in human detection and activity recognition missions is a relatively new topic. Some studies focused on human detection methods from aerial videos in relation to search and rescue missions [11, 12]. These studies aimed at identifying humans lying or sitting on the ground. Some notable approaches related to human identity recognition in low-resolution aerial videos are weighted voter-candidate formulation by Oreifej et al. [13] and blob matching using an adaptive reference set by Yeh et al. [14]. Monajjemi et al. [15] developed a UAV onboard gesture recognition system to identify periodic movements of waving hands from other periodic movements like walking and running in an outdoor environment. Our experimental set-up is most similar to Monajjemi et al.'s.

3. Methodology

This section provides details on human detection, perspective correction, segmentation and feature extraction, pose estimation and trajectory estimation.

3.1. Human detection

A human detection model is trained using the R-CNN method originally presented in [1]. R-CNN combines region proposals with Convolutional Neural Networks (CNN). In the pre-processing stage, it uses a region proposal algorithm [16] before running the CNN. R-CNN is considered to be a state-of-the-art visual object detection system that combines bottom-up region proposals with rich features computed by a CNN [1].

For experimentation, the CNN features of 510 selected images were used to train the detector. 510 images were selected to represent different human subjects from a range of viewpoints. The images were selected from publicly available MoBo Aligned dataset [17], VIRAT video dataset [18], mini-drone video dataset [19], UCF aerial action dataset [20], PETS 2006 dataset [21] and our aerial field videos (see Figure 1). All the human instances in the images were labeled. The images were randomly indexed in order to mix them properly. Then, the dataset was randomly stratified as 0.8:0.2 for training and testing data respectively and achieved a 98% accuracy for human detection.

Transfer learning [22] was applied to retraining the AlexNet [23] neural network with the new CNN features. For this task, AlexNet pre-trained network was selected because it has been pre-trained on 1.2 million ImageNet [24] images of 1000 classes, some of which were trained on images of humans in different settings, and showed the best performance in the ImageNet Large Scale Visual Recognition Challenge in 2012 [23].

3.2. Perspective correction

The relative orientation between the human subject and the camera can be represented in a *horizontal coordinate system* (see Figure 2(a)). In the horizontal coordinate system, $\phi \in [0, \pi/2]$ is the elevation/tilt angle, whereas $\theta \in [0, 2\pi)$ is the azimuth/pan angle. The azimuth angle is calculated in the radial direction between the heading direction of the human subject and the camera center axis on the horizontal plane. The vertical perspective distortion occurs when $\phi > 0$, and worsens as ϕ gets larger. When $\phi = 90°$, perspective distortion cannot be corrected. For $60° \leq \phi < 90°$, the captured images have a severely distorted perspective that is very difficult to compensate. Therefore in this study, the maximum ϕ is limited to 60°.

Perspective correction is done by mapping the distorted image plane (see Figure 2(b)) to the undistorted vertical plane through *homography*. Segments on the undistorted vertical plane then enable the matching of test and training images. Given an image, for every homogeneous point on the image plane, $\boldsymbol{x}$, there exists a homography matrix $\mathbf{H}$ that maps it to a homogeneous point, $\boldsymbol{x}'$, on the undistorted vertical plane, i.e.,

$$\boldsymbol{x}' = \mathbf{H}\boldsymbol{x}. \tag{1}$$

The matrix $\mathbf{H}$ depends on the elevation angle ϕ. Instead of calculating $\mathbf{H}$ for each video, It was calculated offline for each of the following ϕ values: $\arctan(10/30) = 18.4°, \arctan(20/30) = 33.7°, \arctan(30/30) = 45.0°$ and $\arctan(40/30) = 53.1°$. To calculate $\mathbf{H}$, four points were manually selected in a sample video frame to (i) delineate the area of interest and (ii) generate the vertical scene lines, as shown in Figure 2(b). The vertical scene lines define the homography matrix $\mathbf{H}$.

3.3. Segmentation and feature extraction

After perspective correction, the human silhouette was segmented. The size of the silhouette in the image plane varies depending on the direct distance between the camera and the human subject. Perspective correction alone cannot address this scaling issue. Thus, the test silhouette is scaled up or down to match the scale of the training images. Prior to feature extraction, the test videos are annotated for pose and viewpoint.

Figure 1. Sample images with annotated bounding boxes. First three rows represent some images from our field videos. Third to sixth rows correspond to some selected images from UCF aerial action [20], MoBo Aligned [17] and mini-drone [19] datasets respectively.

For each frame, the RGB image was converted into a binary image and its bounding box area was segmented. Noise was removed using a Gaussian filter and small objects containing fewer than a threshold number of pixels were also removed. The remaining blob or blobs were considered to represent the human silhouette. Currently, the denoising parameters and segmentation parameters were customized for each video clip to obtain the best possible silhouette, so they are subject to improvements.

For feature extraction, the image window was divided into small spatial regions called "HOG cells" [2]. The weighted gradients in a HOG cell form a 1-D histogram which represented the orientation of the edge lines. The feature vector was formed from the HOG blocks, each of which represents a group of HOG cells.

3.4. Pose estimation

A training dataset was created from 1017 silhouette images to identify the eight sub-steps of the human gait cycle (P_1 to P_8 in Figure 3 (b)). This training dataset should not be confused with the 510 images used to train the R-CNN detector. Each sub-step (or pose) had viewpoints from eight radial directions (azimuth angles that are 45° apart), giving rise to $8 \times 8 = 64$ pose-viewpoint pairs. The finite number of elevation-azimuth angle pairs are equivalent to the

IOP Conf. Series: Materials Science and Engineering **405** (2018) 012003 doi:10.1088/1757-899X/405/1/012003

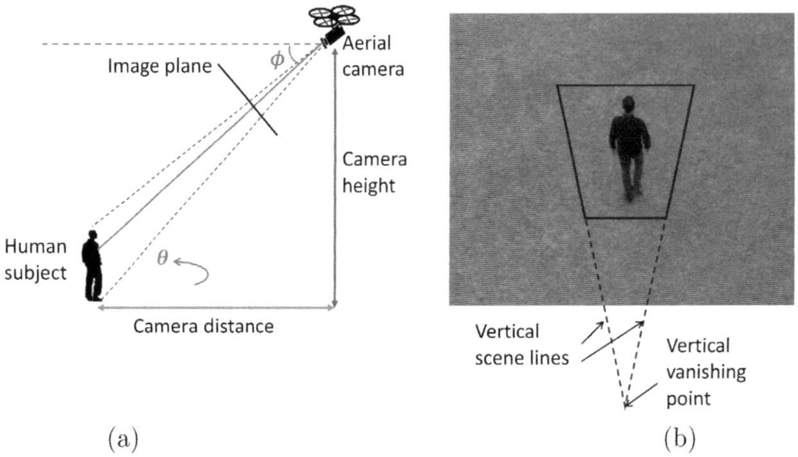

Figure 2. (a) The UAV hovers at a known camera height and angle. (b) An input image with vertical scene lines. The scene lines are manually constructed according to the elevation angle. The blue color box in the middle is the area of interest for homography in the vertical plane. The vanishing point is the point where parallel scene lines would meet each other on the image plane.

discretized viewing hemisphere described in [6].

The collected training data consisted of 64 labels, representing eight sub-steps of the gait cycle and eight viewpoints (see Figure 3 (b)). A training dataset of n observations is denoted by

$$S = \{(\boldsymbol{x}_1, y_1), (\boldsymbol{x}_2, y_2), \ldots, (\boldsymbol{x}_n, y_n)\}, \tag{2}$$

where $\boldsymbol{x}_i$ is the ith feature vector, and y_i the ith label.

It was assumed that the human subject walked forward at a constant speed, does not take sharp turns and does not twist the body while turning. This assumption does not preclude left, right, or backward turns, as long as the turning is not abrupt, as exemplified by the yellow-border windows in Figure 3(b).

The classification output was considered to be a state. The admissible state transitions restrict the next classifier prediction to be one of the states the current state can transition to (yellow-border windows in Figure 3(b)). Given the current pose and viewpoint, when a new image is available, the associated pose was predicted to be either the current pose, or the pose in the next sub-step of the gait cycle. When the pose changes from the current state to the next state, the viewpoint of the next pose has to be one of the following: the same viewpoint (moving straight), 45° clockwise from the current viewpoint (turning left), or 45° anticlockwise from the current viewpoint (turning right).

Our dynamic classifier selection (DCS) architecture consists of 64 4-class SVM classifiers denoted $C_4(P_i, V_j)$, $i, j \in \{1, \ldots, 8\}$. The classifier $C_4(P_i, V_j)$ is associated with pose P_i and viewpoint V_j, and is trained to recognize the set of four classes:

$$\{(P_i, V_j), (P_{i\boxplus 1}, V_j), (P_{i\boxplus 1}, V_{j\boxminus 1}), (P_{i\boxplus 1}, V_{j\boxplus 1})\}, \tag{3}$$

where $i, j \in \{1, \ldots, 8\}$ and the operators $\boxplus, \boxminus$ are defined as follows:

$$i \boxplus j = (i + j + 1) \mod 8 - 1, \tag{4}$$
$$i \boxminus j = (i - j - 1) \mod 8 + 1. \tag{5}$$

For example, the classifier $C_4(P_4, V_5)$ is trained to recognize the four classes labeled a, b, c and d in Figure 3 (b).

As depicted in Figure 3 (a), our classification process works in two stages: (i) the initialization stage and (ii) the DCS stage. In the initialization stage, the first q video frames are classified using classifier C_{64}. C_{64} is a single 64-class SVM classifier trained with the complete dataset (64 classes). The DCS stage starts with the $(q + 1)$th video frame. In this stage, each frame is classified with a classifier chosen based on the class label predicted by the previous iteration.

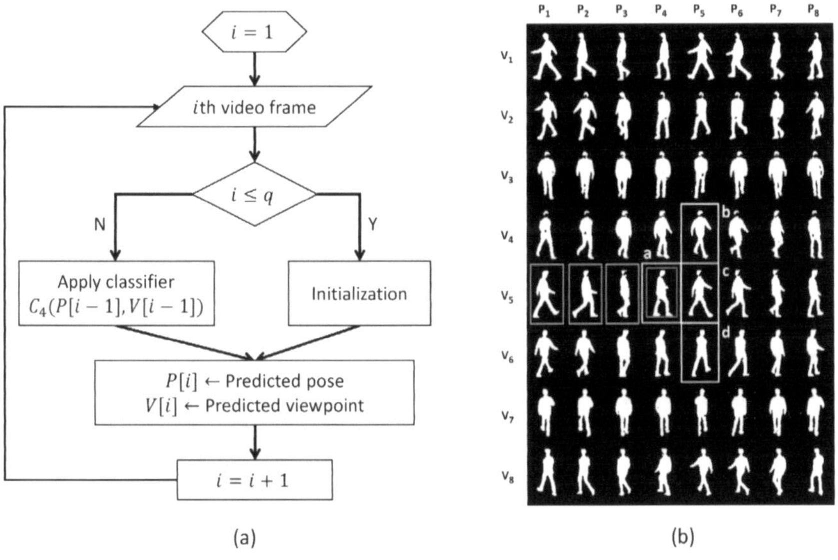

Figure 3. (a) Flowchart for pose-viewpoint classification by dynamic classifier selection. Prior to the workflow, all classifiers should have been trained. (b) The training dataset consists of 64 classes of pose-viewpoint pairs. The rows and columns represent the poses and the viewpoints respectively. Each silhouette in the figure is a random image from the pose-viewpoint subset it belongs. Blue- and red-border windows show four consecutive initialized frames (from left to right). Once initialized, the pose and the viewpoint of the most recently initialized image (red-border window) are used to select the next classifier. In this example, the training image subsets of the next classifier are shown in yellow-border windows.

For elaboration, consider the example in Figure 3(b). Suppose $q = 4$, and the blue- and red-border windows are sample classes predicated by the classifier C_{64}. The red-border window highlights the class predicted for the qth frame. Since this class is (P_4, V_5), the classifier $C_4(P_4, V_5)$ is chosen to classify the $(q + 1)$th frame. The training subsets for $C_4(P_4, V_5)$ are highlighted with the yellow-border windows a, b, c and d.

The most significant difference between our classifier architecture and existing architectures in the recent literature [3, 25, 26] is that ours does not execute all the classifiers to make a decision. Instead, only the relevant classifier is selected for every next image. The relevance of the classifier is determined by its training subsets, and the training subsets are selected based on the state transition graphs.

3.5. Trajectory estimation

Trajectory estimation refers to the estimation of the shape of the path traversed by the human subject. Each estimated viewpoint serves as an estimation of the walker's orientation. For each estimated orientation, a 3-D pose is reconstructed from the estimated pose. The algorithm can be described as follows:

- Whenever an estimated pose is the same as the previous, he subject is assumed to remain at the same location. Such predictions occur due to the camera's high frame rate and/or the subject's slow movements.
- Whenever an estimated pose differs from the previous, the subject is assumed to have moved a fixed distance from the location of the previous pose. When the orientation changes, the next pose is positioned at a fixed distance from the location of the previous pose at an angle of $\pm 45°$ (+ve for right turns, −ve for left turns).

4. Experimental results

The experiments were conducted at different heights with original aerial videos and perspective-corrected videos. For trajectory estimation, each estimated trajectory is plotted on a 2-D plane with *unitless* axes, and the starting location mapped to the origin. Along a trajectory, the estimated poses were reconstructed using a 3-D, 13-jointed skeletal models. The proximty of the estimated trajectories to the actual trajectories was assessed visually.

All the videos were captured from a rotorcraft UAV (see Figure 4(b)) in a slow and low-altitude flight mode. For recording videos, a GoPro Hero 4 black camera with an anti-fish eye replacement lens (5.4mm, 10MP, IR CUT) and a 3-axis Solo gimbal was used. The images were sampled at a rate of 30fps. In order to ease the segmentation process, the videos were recorded with an uncluttered background and with the human subject wearing dark clothes. The UAV-captured videos were segmented as described in Section 3.3. These experiments were conducted using HOG features.

Certain assumptions were made to ease the coordinate transformation between the camera and the human subject. The human subject was assumed to be upright on a flat ground. The camera roll angle was considered to be zero. The roll, pitch and yaw angles of the UAV were assumed to be zero during the slow flying. Hence, the flight dynamics of the UAV has negligible effects on the true camera elevation angle. The camera elevation angle and the height were directly recorded via the UAV control interface. The UAV was operated at a known ground distance (camera distance) from the human subject.

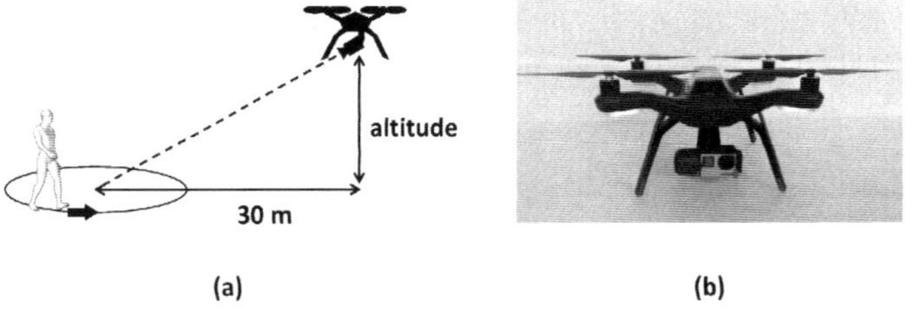

Figure 4. (a) A human subject was filmed walking on a circle, while the UAV stays pointed at the center of the circle. (b) The rotorcraft UAV, namely a 3DR Solo, used for experimentation.

As depicted in Figure 4(a), a human subject is filmed walking on a marked circle by a UAV pointing at the center of the circle. The experiment was conducted to analyze the effect of perspective distortion in detail. The UAV was flown at heights of 10m, 20m, 30m and 40m (see Figure 6). The lowest height of 10m caused negligible perspective distortion, but at $h = 40$m ($\phi = 53.1°$), the video suffers from severe perspective distortion. The main observations are:

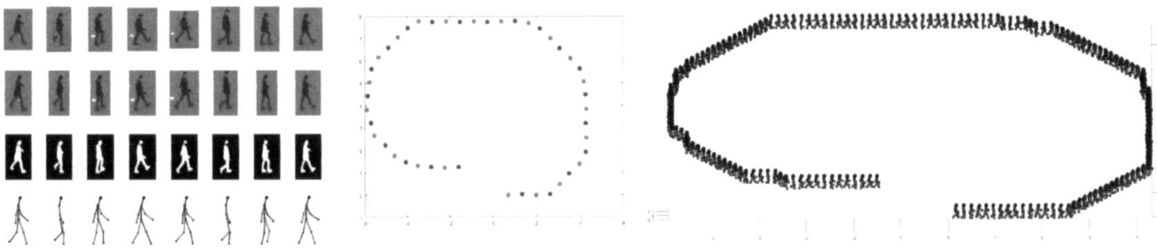

Figure 5. Results for the altitude of 20m: (a) The top row shows a series of cropped video frames. The second row is the perspective-corrected version of the top row. The third row shows the segmented silhouettes and the bottom row shows the estimated poses. (b) The estimated trajectory, where each dot marks where both feet touch the ground, with red representing right foot in front and blue representing left foot in front. (c) The estimated trajectory and 3-D reconstruction of the estimated poses.

Table 1. Estimation errors of the dynamic classifier for perspective-distorted (PD) and perspective-corrected (PC) videos. e_{pose} and $e_{\text{viewpoint}}$ are estimation errors for pose and viewpoint respectively.

Altitude	Perspective distortion	#frames	e_{pose}	$e_{\text{viewpoint}}$
$h = 10$m	No distortion	787	**23.5%**	**13%**
$h = 20$m	PD	784	**22.1%**	**17.2%**
	PC		39.9%	20.4%
$h = 30$m	PD	810	56.7%	44.8%
	PC		**40.6%**	**37.2%**
$h = 40$m	PD	817	74.4%	42.6%
	PC		**37.3%**	**24.8%**

- In terms of pose and viewpoint estimation accuracies, perspective correction helps the dynamic classifier.
- The advantage of perspective correction is more pronounced on more distorted videos.

Table 1 shows an overall reduction in estimation errors when perspective correction is applied. This conclusion is further indicated in improved trajectory estimations in Figure 6.

5. Discussion

The human detector trained using R-CNN successfully identifies the humans in aerial videos. However, it is trained to focus human detection from relatively clutter-free backgrounds. The error rate might be high for detecting humans in a complex background. The robustness of the detector can be improved by using a range of images from different settings as training set images.

A drawback of the dynamic classifier is its dependency on the initialization. Like all classifiers, C_{64} sometimes makes mistakes, throwing the $C_4(\cdot,\cdot)$ classifiers off-course. A potential improvement is to re-initialize the dynamic classifier (see Figure 3(a)) periodically.

HOG features are traditionally considered to be handcrafted features, and in some areas they have been replaced by CNN features. However, for the classification stage we are interested in

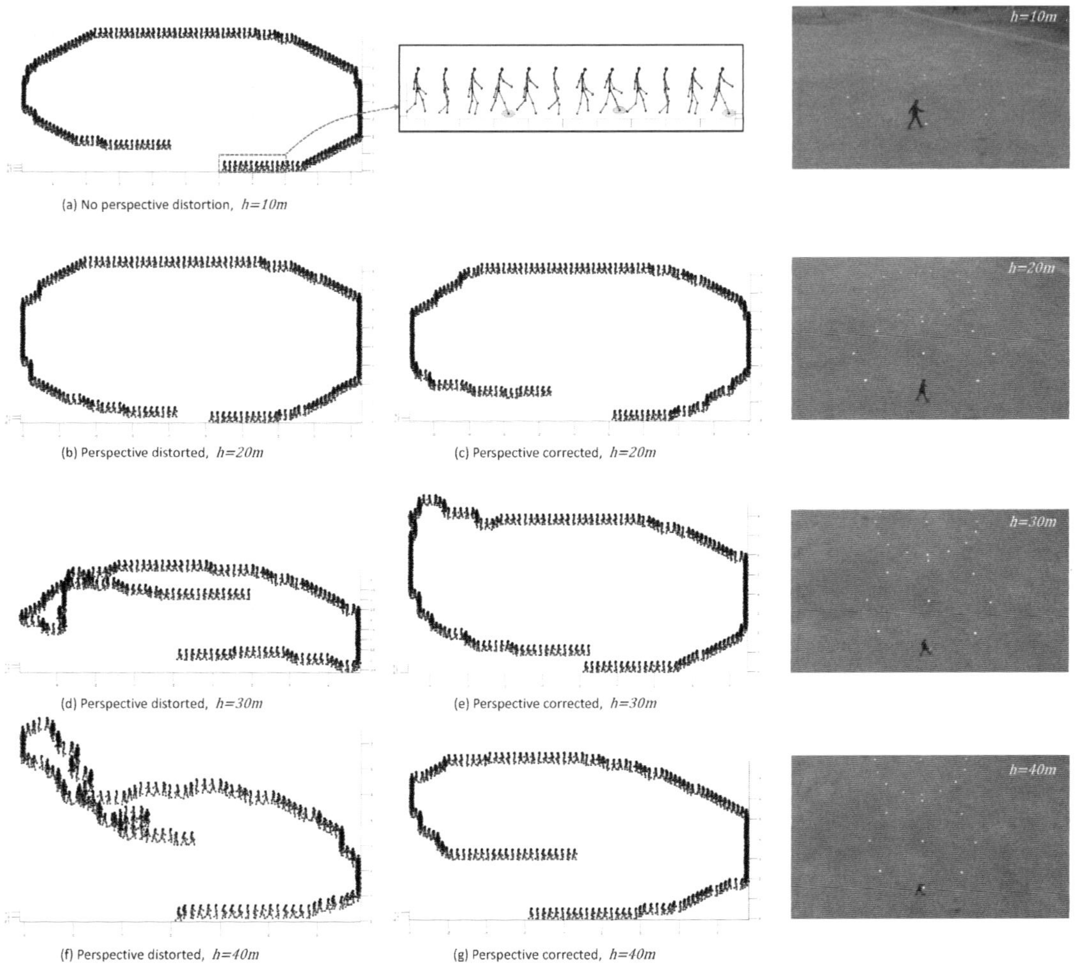

(a) No perspective distortion, *h=10m*

(b) Perspective distorted, *h=20m*

(c) Perspective corrected, *h=20m*

(d) Perspective distorted, *h=30m*

(e) Perspective corrected, *h=30m*

(f) Perspective distorted, *h=40m*

(g) Perspective corrected, *h=40m*

Figure 6. Reconstructed poses and their trajectories correspond to four different heights.

the shapes of the human subject. Our observation for the overall robustness of HOG features is dependent on silhouettes and hence significantly depends on the edges. However, segmentation of aerial images (for HOG) is very challenging due to varying resolution and background, and can benefit from the latest advances in semantic segmentation.

The results confirm the intuition that perspective correction is imperative for severely perspective-distorted videos. Our solution has problems with purely frontal or back views, because frontal and back silhouettes do not provide sufficient details for differentiating poses. A potential solution is provided by the mobility of the aerial platform itself. The UAV can be programmed to seek a good elevation angle and azimuth angle, before it starts analyzing the human subject's action. This will require control algorithms and machine intelligence that go beyond the scope of this work.

6. Conclusion

This paper discusses a solution for human detection from perspective distorted aerial videos and estimates their pose and trajectory. The approach consists of R-CNN-based detection, perspective correction by homography, HOG feature extraction and dynamic classifier selection. The detector trained on different aerial and fronto-parallel images achieved nearly ideal accuracy in the experiments. The dynamic classifier consists of 64 4-class classifiers and enables robust classification results. Trajectory estimation provides the shape of the path traversed by the human subject, and is dependent on viewpoint estimation. The study discussed in this article is limited to estimating walking gaits. Our future work includes equipping UAVs with the ability to recognize human activities.

Acknowledgment
This project was partly supported by Project Tyche, the Trusted Autonomy Initiative of the Defence Science and Technology Group.

References

[1] Girshick R, Donahue J, Darrell T and Malik J 2014 Rich feature hierarchies for accurate object detection and semantic segmentation *2014 IEEE Conference on Computer Vision and Pattern Recognition* pp 580–587 ISSN 1063-6919
[2] Dalal N and Triggs B 2005 Histograms of oriented gradients for human detection *2005 IEEE Computer Society Conference on Computer Vision and Pattern Recognition (CVPR'05)* vol 1 pp 886–893 vol. 1 ISSN 1063-6919
[3] Woods K, Kegelmeyer W P and Bowyer K 1997 Combination of multiple classifiers using local accuracy estimates *IEEE Transactions on Pattern Analysis and Machine Intelligence* **19** 405–410 ISSN 0162-8828
[4] Ko A H, Sabourin R, Britto A S and Jr 2008 From dynamic classifier selection to dynamic ensemble selection *Pattern Recognition* **41** 1718 – 1731 ISSN 0031-3203
[5] Hartley R and Zisserman A 2003 *Multiple view geometry in computer vision* (Cambridge university press)
[6] Rogez G, Orrite C, Guerrero J and Torr P H 2014 Exploiting projective geometry for view-invariant monocular human motion analysis in man-made environments *Computer Vision and Image Understanding* **120** 126 – 140 ISSN 1077-3142
[7] Bak S, Zaidenberg S, Boulay B and Brmond F 2014 Improving person re-identification by viewpoint cues *2014 11th IEEE International Conference on Advanced Video and Signal Based Surveillance (AVSS)* pp 175–180
[8] Li Y, Wu B and Nevatia R 2008 Human detection by searching in 3d space using camera and scene knowledge *2008 19th International Conference on Pattern Recognition* pp 1–5 ISSN 1051-4651
[9] 2004 URL `http://homepages.inf.ed.ac.uk/rbf/CAVIARDATA1/`
[10] Perera A G, Law Y W, Al-Naji A and Chahl J 0 2018 Human motion analysis from UAV video *International Journal of Intelligent Unmanned Systems* **0** 00–00 (*Preprint* `https://doi.org/10.1108/IJIUS-10-2017-0012`) URL `https://doi.org/10.1108/IJIUS-10-2017-0012`
[11] Rudol P and Doherty P 2008 Human body detection and geolocalization for uav search and rescue missions using color and thermal imagery *Aerospace Conference, 2008 IEEE* pp 1–8 ISSN 1095-323X
[12] Andriluka M, Schnitzspan P, Meyer J, Kohlbrecher S, Petersen K, von Stryk O, Roth S and Schiele B 2010 Vision based victim detection from unmanned aerial vehicles *Intelligent Robots and Systems (IROS), 2010 IEEE/RSJ International Conference on* pp 1740–1747 ISSN 2153-0858
[13] Oreifej O, Mehran R and Shah M 2010 Human identity recognition in aerial images *Computer Vision and Pattern Recognition (CVPR), 2010 IEEE Conference on* pp 709–716 ISSN 1063-6919
[14] Yeh M C, Chiu H K and Wang J S 2016 Fast medium-scale multiperson identification in aerial videos *Multimedia Tools and Applications* **75** 16117–16133 ISSN 1573-7721
[15] Monajjemi M, Bruce J, Sadat S A, Wawerla J and Vaughan R 2015 UAV, do you see me? Establishing mutual attention between an uninstrumented human and an outdoor UAV in flight *2015 IEEE/RSJ International Conference on Intelligent Robots and Systems (IROS)* pp 3614–3620
[16] Zitnick C L and Dollár P 2014 Edge boxes: Locating object proposals from edges *Computer Vision – ECCV 2014* ed Fleet D, Pajdla T, Schiele B and Tuytelaars T (Cham: Springer International Publishing) pp 391–405
[17] Rogez G, Rihan J, Orrite-Uruñuela C and Torr P H S 2012 Fast Human Pose Detection Using Randomized Hierarchical Cascades of Rejectors *International Journal of Computer Vision* **99** 25–52 ISSN 1573-1405 URL `https://doi.org/10.1007/s11263-012-0516-9`
[18] Oh S, Hoogs A, Perera A, Cuntoor N, Chen C C, Lee J T, Mukherjee S, Aggarwal J K, Lee H, Davis L, Swears E, Wang X, Ji Q, Reddy K, Shah M, Vondrick C, Pirsiavash H, Ramanan D, Yuen J, Torralba

A, Song B, Fong A, Roy-Chowdhury A and Desai M 2011 A large-scale benchmark dataset for event recognition in surveillance video *CVPR 2011* pp 3153–3160 ISSN 1063-6919

[19] Bonetto M, Korshunov P, Ramponi G and Ebrahimi T 2015 Privacy in mini-drone based video surveillance *2015 11th IEEE International Conference and Workshops on Automatic Face and Gesture Recognition (FG)* vol 04 pp 1–6

[20] 2009 UCF Aerial Action Data Set `http://crcv.ucf.edu/data/UCF_Aerial_Action.php` [Online; accessed 19-Jan-2018]

[21] 2006 PETS 2006 Benchmark Data `http://www.cvg.reading.ac.uk/PETS2006/data.html` [Online; accessed 19-Jan-2018]

[22] Yosinski J, Clune J, Bengio Y and Lipson H 2014 How transferable are features in deep neural networks? *Advances in Neural Information Processing Systems 27* ed Ghahramani Z, Welling M, Cortes C, Lawrence N D and Weinberger K Q (Curran Associates, Inc.) pp 3320–3328

[23] Krizhevsky A, Sutskever I and Hinton G E 2012 Imagenet classification with deep convolutional neural networks *Advances in Neural Information Processing Systems 25* ed Pereira F, Burges C J C, Bottou L and Weinberger K Q (Curran Associates, Inc.) pp 1097–1105 URL `http://papers.nips.cc/paper/4824-imagenet-classification-with-deep-convolutional-neural-networks.pdf`

[24] Deng J, Dong W, Socher R, Li L J, Li K and Fei-Fei L 2009 Imagenet: A large-scale hierarchical image database *2009 IEEE Conference on Computer Vision and Pattern Recognition* pp 248–255 ISSN 1063-6919

[25] Kuncheva L I, Bezdek J C and Duin R P 2001 Decision templates for multiple classifier fusion: an experimental comparison *Pattern Recognition* **34** 299 – 314 ISSN 0031-3203

[26] Tulyakov S, Jaeger S, Govindaraju V and Doermann D 2008 *Review of Classifier Combination Methods* (Berlin, Heidelberg: Springer Berlin Heidelberg) pp 361–386 ISBN 978-3-540-76280-5

Controller response behaviour during procedural control with surveillance information

S. M. B. Abdul Rahman[1,*], M. F. Sidik[2], M. H. I. Mohd Shukri[1], M. H. Nazarudin[1]

[1] Flight Technology & Test Centre, Faculty of Mechanical Engineering, Universiti Teknologi MARA, 40450 Shah Alam, Selangor, Malaysia
[2] Kuala Lumpur Air Traffic Control Centre, Department of Civil Aviation Malaysia, Subang, Malaysia.

mariam4528@salam.uitm.edu.my

Abstract. Air Traffic Controller (ATCo) duties are challenging due to their heavy task demand. ATCo needs to monitor air traffic while ensuring a smooth flow of traffic and also maintaining safe separation between aircraft. In this study, a human-in-the-loop experiment was carried out to assess the controller response behaviour during procedural control with radar information assistance. Lumpur Sector 4 in Kuala Lumpur Flight Information Region (KL FIR) has been chosen as the assigned sector to study ATCo reactions during control activities. The sector was characterised with two intercepting inbound flight routes and two sets of traffic feed: high traffic feed with incoming feed rate of one aircraft for every 60s and low traffic feed with incoming feed rate of one aircraft for every 70s. The experiment was conducted with two controllers (expert subjects) from the Department of Civil Aviation (DCA) and two students (trained subjects) using a customized air traffic radar and either a Procedural Control Bay or an Electronic Flight Strip Bay. To monitor controller reactions, a heart rate sensor monitoring device was chosen as a primary monitoring tool. Additionally, the controller subjective workload was measured using the Instantaneous Self-Assessment (ISA) workload rating and NASA Task Load Index (NASA-TLX) questionnaires. Based on the results, the ISA workload rating can be observed to be fluctuating throughout the experiment sessions. The fluctuation was also captured to be in agreement with the response gathered from heart rate sensor monitoring device. Overall, the subjects were observed to exhibit a lower stress level for the low traffic condition than for the high traffic condition. In the future, it is recommended that more subjects should be included in the research in order to be better conclude on whether workload can indeed be assessed by monitoring subject's heart rate or response.

1. Introduction

One of the key challenges of Air Traffic Controller (ATCo) tasks is to ensure safety of flight crews and passengers onboard through safe and efficient flow of air traffic [1]. They have to maintain situation awareness in condition of rapidly changing information [1]. ATCo have the responsibility to decide with conflicting goals and high responsibility under tight time constraints, which causes them a tremendous cognitive workload [1]. Important and quick decisions have to be made such as deciding on suitable flight level, speed and heading to prevent future separation minima infringement between aircraft.

These decision making skills are included in cognitive complexity of a controller, which describes as the relationship between the controller tasks to handle traffic and the corresponding mental workload [2]. Previous studies on cognitive analysis have demonstrated that controller duties are related to high pressure while monitoring traffic and interacting with pilots [3]. Consequently, unwanted consequences may occur due to the controller's error. Human errors may occur especially during interactions between the pilots and the controllers, which are most likely to threaten the flight safety [4].

1.1. Air traffic control
Air Traffic Control (ATC) is the service provided by ATCo, who is responsible for assisting, dispatching and maintaining a secure, safe and systematic flow of air traffic [3]. Generally, there are three different levels of air traffic operations that correlate to the major phases of flight: (i) aerodrome control for take-off, landing and ground movement of aircrafts, (ii) approach control for arriving and departing aircrafts, and (iii) area control based in the Air Traffic Control Center (ATCC) for handling the en-route phase of flight [5]. This research only focuses on the duties of controller in the en-route phase that also known as area control. En-route controllers resume control of aircraft from approach controller and are required to maintain adequate separation from other traffic during the climbing, descending and cruising phases through projection of aircraft position and trajectory in time and space [6].

In ATC system, surveillance and procedural control are two commonly used control strategies in managing traffic movement within a specified airspace. Procedural control method depends largely on the ATCo's mental capacity where they need to imagine the position and the trajectory of an aircraft. Future conflicting traffic shall be estimated manually. To help with this process, a Flight Progress Strip is commonly used to continuously update aircraft position and movement. Flight Progress Strip used in Malaysia consists of twenty-four fields' information that is printed on a rectangular-shape paper strip as shown in Figure 1. Flight Progress Strip provides essential information of an aircraft such as call sign, planned route, aircraft type, filed airspeed, assigned altitude, estimated time over certain waypoints and other relevant information [7]. Procedural control and surveillance control can co-exist together where procedural control is used to determine future conflict and surveillance control is used to resolve on time conflict. However, in certain geometrical area or circumstances where the surveillance coverage is not available, ATCo is provided with limited option in controlling traffic movement.

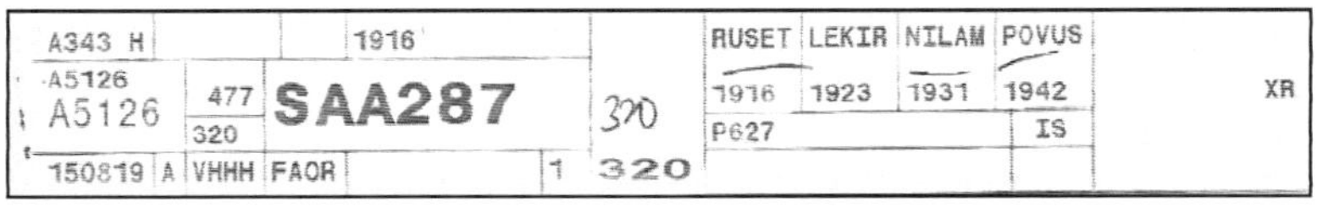

Figure 1. Physical Flight Progress Strip

1.2. Human performance measures
Evaluation of ATCo performance can be investigated under several concepts such as human behavior rather than human workload, and measurement techniques, which are subjective rather than objective, or scientifically in physiological rather than psychological, in order to obtain better result for the research [8]. The objective of this research is to assess controller response behavior in ATC during procedural control activities by manipulating few measurable air traffic factors that affects the mental capacity of a controller. Psychophysiological measures may prove especially useful in the prevention of performance deterioration in under load or overload conditions [9]. By this definition, when the workload is low, very little cognitive effort is required and the operator can easily complete the task. On the other extreme, when workload is very high, much cognitive effort is required that an operator is likely to leave the task unfinished [10]. Many studies implicitly assume that the controller workload varies as a function of both directly measurable air traffic factors (number of aircraft in the sector, speed variability, proximity of aircraft, etc.) and controller's activity mediated by factors such as the controller's abilities, age, fatigue, and level of experience [11].

From research by Lee, Pathirana and Caelli (2013), they have demonstrated the efficiency of Doppler Radar of 2.7 GHz operating band in analyzing breathing patterns under various breathing forms like normal breathing, rapid and slow breathing, as well as different rates of inhale and exhale [12]. In order to assess the mental state of a person for this research, the measurement of heart rate (HR) and pulse rate variability (PRV) is seen as a very powerful monitoring tool. Even though both measurement tools have yet to be conducted on the ATCo, a physiological measurement always seems as a suitable method for the evaluation of human factor. To measure the controller response behavior, a human-in-the-loop (HITL) experiment was conducted in a prototype environment of ATCC with a main purpose to evaluate the concept in terms of practicability, controller workload impact and potential benefits [13]. The HITL experiment in this research was presented through the mock-up environment of ATC. The experiment was performed by two controllers from Kuala Lumpur ATCC as expert subjects and by two students as trained subjects.

In this research, workload is measured using three different measurements through subjective ratings and behavioural/physiological recordings. The evaluation of controller situation awareness and response behavior was measured using a heart rate sensor device, whilst the evaluation of ATCo subjective ratings was made using Instantaneous Self-Assessment (ISA) workload rating together with questionnaire tool from NASA Task Load Index (TLX). The ISA workload rating tool was intended for ATCo to report their perceived workload during experiments or simulations [14]. In this experiment, the controller was prompted to give a rating from 1 to 7 of their ISA workload level at every 1 minute interval during the HITL experiment. The NASA TLX, on the other hand, is a comprehensive multi-dimensional subjective task load rating technique [15], which consist of six dimensions of the subject's task load experiences: mental demand, physical demand, temporal demand, perceived performance, efforts and frustration level. NASA TLX has been applied extensively in similar experimental environment. The NASA TLX survey was given at the end of each session for this HITL experiment.

2. Methodology

2.1. Kuala Lumpur Flight Information Region

Flight Information Region (FIR) is the airspace of defined dimensions, which flight information service and alerting service are provided. Both services are the basic air traffic services to provide information pertinent to the safe and efficient conduct of flights and in alerting the relevant authorities for aircraft in need. KLFIR is divided into six geographical sectors in order to have a systematic framework for ATC. Based on the six sectors in KLFIR, Lumpur Sector 4 was chosen as the experimental field for this research as it geometrically located in oceanic area and partially covered by surveillance. Lumpur Sector 4 is located in between of Phuket in Thailand and Banda Aceh, Indonesia. It has a total of 7 RNAV route with 7 crossing waypoint in about 300 nautical miles range. However, in order to have a more controlled scenario, only two main inbound routes with one merging point were simulated in this experiment as shown on Figure 2.

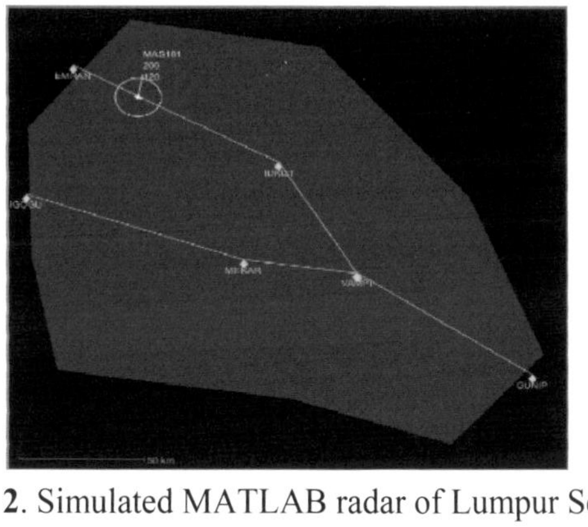

Figure 2. Simulated MATLAB radar of Lumpur Sector 4

The level of traffic difficulties is then simulated with different levels of incoming traffic feed, namely high traffic with incoming traffic rate of 60 seconds per aircraft and low traffic with incoming traffic rate of 70 seconds per aircraft.

2.2. Experiment settings

Two expert subjects and two trained subjects volunteered for the experiment. The expert subjects are from KLATCC, both aged 30 and with 5 years of experience. The trained subjects are from final year students of Mechanical Engineering Faculty that was comprehensively briefed on the ATC tasks and working manner. The experiment used two distinct subject groups in order to assess whether the use of heart rate monitor is relevant for different sets of user's background.

The experiment began with a briefing session explaining the experiment's objective together with explaining the course of the session, followed by a training session/s for each subject to familiarize with the experiment settings. Each scenario was conducted for 15 minutes. The scenario was divided into two types of traffic feed, which were high traffic and low traffic scenarios. High traffic scenario has a maximum of 10 aircraft at one time and an incoming feed rate of 60s per aircraft, while the low traffic scenario has a maximum number of 8 aircraft at one time and an incoming feed rate of 70s per aircraft. All aircraft have been arranged to fly randomly into the sector from either EMRAN or IGOGU. The experiment was conducted using a customized surveillance display from Matlab® together with either Procedural Control Bay or Electronic Flight Strip Bay. The experiment for all subjects was arranged as shown in Table 1.

Table 1. Experiment flow

	1		2		3		4	
	Traffic	**Control Panel**	**Traffic**	**Control Panel**	**Traffic**	**Control Panel**	**Traffic**	**Control Panel**
	Briefing and Training Session							
1	High	PCB	High	EFSB	High	PCB	High	EFSB
2	Low	PCB	Low	EFSB	Low	PCB	Low	EFSB
3	Low	EFSB	Low	PCB	Low	EFSB	Low	PCB
4	High	EFSB	High	PCB	High	EFSB	High	PCB

PCB: Procedural Control Bay
EFS: Electronic Flight Strip Bay

Figure 3 and 4 illustrate a complete setup for the HITL experiment. During the experiment, subject was placed in the Controller Station, while 3 pseudo-pilots were placed in the Pseudo-Pilot Station. They were separated by 15 meters between them in order to simulate the long-range communication between controller and pilot.

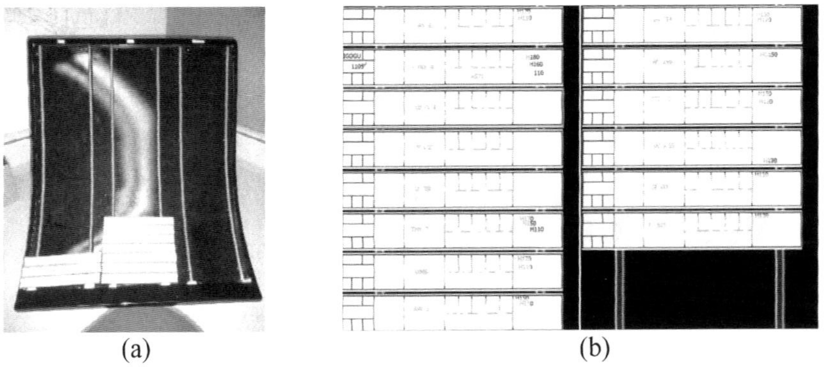

(a) (b)

Figure 3. Control panel options: (a) Procedural control bay, (b) flight strip manager software

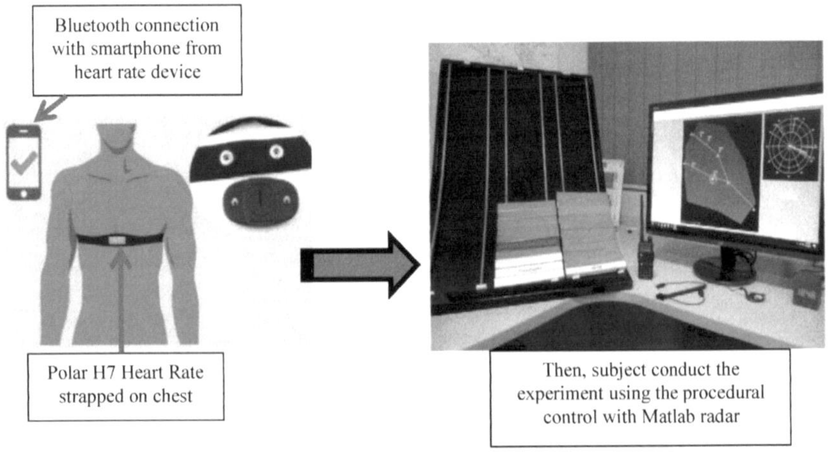

Figure 4. Illustration of equipment setup

3. Results and discussion

The evaluation of controller situation awareness and response behavior was measured using a heart rate sensor device, whilst the evaluation of ATCo subjective ratings was made using ISA workload rating together with questionnaire tool from NASA TLX. The experiment also looked into the resulting traffic pattern, especially occurrences of minimum separation infringement or referred to as conflict hereafter. The minimum separation between aircraft was set at 5 nautical miles and any aircraft that fly within the separation limit will be regarded as conflicting traffic.

3.1. Expert subject

Figures 5 and 6 show heart rate readings together with ISA workload rating for both high and low traffic, respectively, for Subject 1 (expert) while using the Procedural Control Bay. Based on the response from Subject 1, the conflicts in Figure 5 occured because of the same level assignment of aircraft based from the scenario given. In common practice, if this scenario happened, ATCo will reassign a new cruising altitude for the aircraft in order to maintain separation between intercepting traffic. However, this was not possible for the scenario given as the subjects were only allowed to give flight speed and heading changes in order to ensure safe separation between aircraft. For conflict point number 2, the heart rate reading is seen to increase drastically as the subject encountered conflict situation upon hand-over to the neighboring sector (Figure 7 (b)). The increase in workload is also confirmed through the subject's ISA workload rating given during the conflict period.

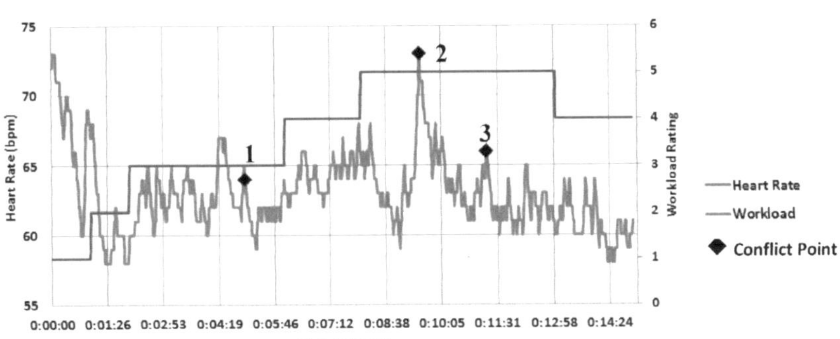

Figure 5. High traffic for Subject 1 using Procedural Control Bay

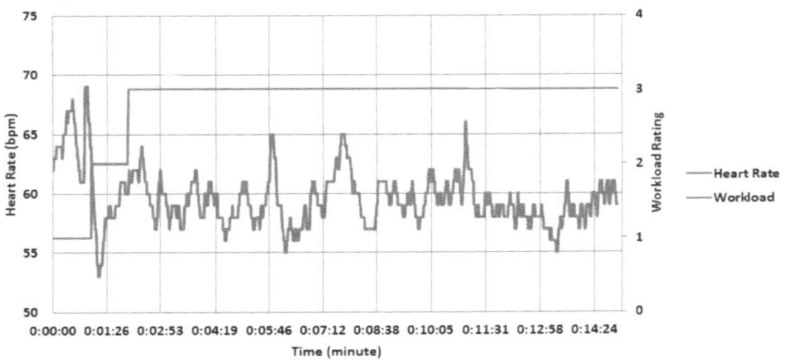

Figure 6. Low traffic for Subject 1 using Procedural Control Bay

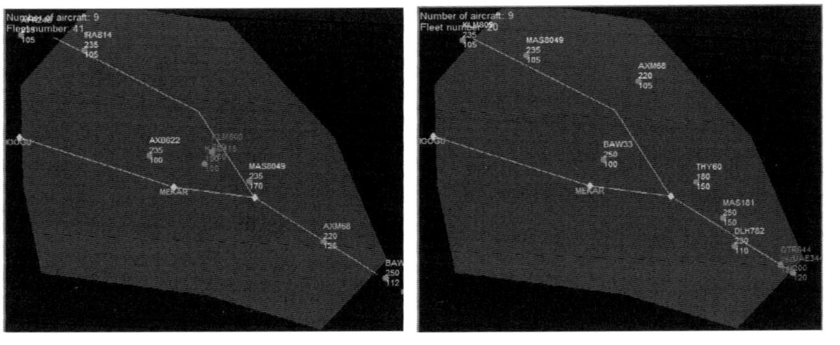

(a) Conflict 1 (b) Conflict 2

Figure 7. Conflicts from Subject 1 using Procedural Control Bay

Furthermore, Subject 1 highlighted that the high traffic feed setting that has a maximum number of aircraft of 10 was giving a higher than expected demand to the subject without leaving much parameter to control the traffic since he was only allowed to use flight speed and heading. However, in low traffic feed scenario, the subject was able to control the traffic while having zero conflict with a consistently low workload rating throughout the 15 minutes-scenario. Having said that, it should be made aware that the use of heading command is not a common standard procedure during procedural control at Lumpur Sector 4. With both expert subjects from KLATCC who are an area procedural and surveillance rated ATCo, they were already briefed and were clear on the control options available during the experiment.

Ideally, participants are required to provide a separation between the aircraft from both airways by controlling their time of arrival at merging waypoint. However, during the exercise when subjects were unable to control the situation and the method of controlling aircraft time of arrival at merging point was not feasible, for traffic from MEKAR eastward, participants were allowed to use surveillance control methods whereby they were allowed to change the aircraft's flight path by assigning heading in order to maintain adequate separation before handing over to the next sector, which is on surveillance control environment.

For simulation using Electronic Flight Strip Bay, conflicts only occurred during high-load condition as shown on Figure 8. From the graph, conflict occurrences is represented by the three diamond points, which indicate three conflicts occurrences for Subject 1. The red colored aircraft on the radar screen in Figure 9 represents conflict detection while the green colored aircraft represents aircraft that has been hand over to the other sector. The first and third conflict occurred between minutes 5 to 6 and 11 to 12. During these instances, the conflict occurred when Subject 1 already transferred the aircraft to the other sector (Figure 9 (a)). Therefore, the heart rate reading of Subject 1 is not in accordance to the reported workload rating. Meanwhile, when the second conflict occurred between minutes 9 to 10, the heart rate of Subject 1 had suddenly increased due to the conflict that occurred near the intercepting point (Figure 9 (b)).

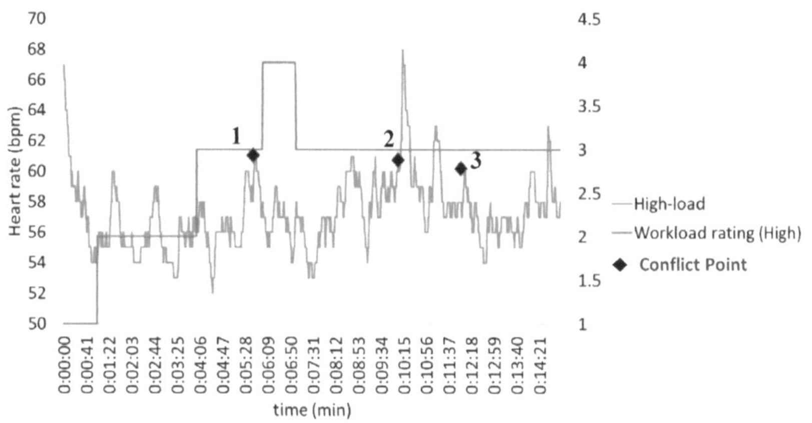

Figure 8. High traffic for Subject 1 using Electronic Flight Strip Bay

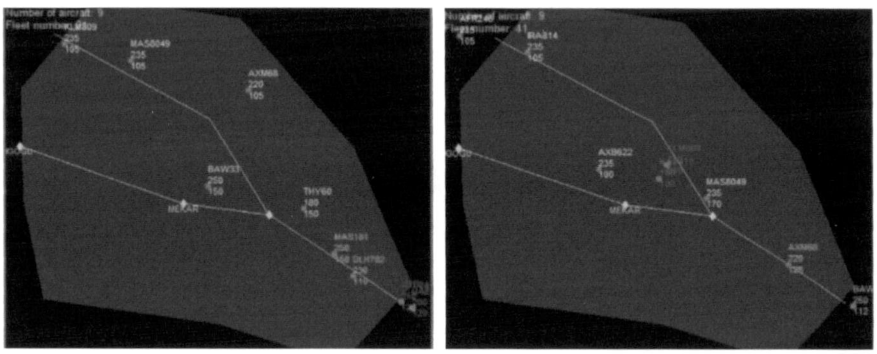

(a) Conflict 1 (b) Conflict 2

Figure 9. Conflicts from Subject 1 using Electronic Flight Strip Bay

In ATC, the strategies selection and execution have a big influence on the performance of an ATCo. Strategies used depend on individual skills and experiences. Based on Figure 10 (a), Subject 1 was not actively using the Procedural Control Bay as one of the tools to control air traffic during the experiment. Hence the subject was unable to perform very well without causing any conflicts during the experiment. The subject needs to invest more time to track the correct strip for the aircraft in the radar during his experiment session. As a result, Subject 1 recorded a total of three conflicts during high traffic scenario as shown previously in Figure 5. Figure 10 (b) shows a better strategy practiced by the Subject 2. He used Procedural Control Bay to arrange and manage the flight strips efficiently. Subject 2 only recorded one conflict in high traffic scenario.

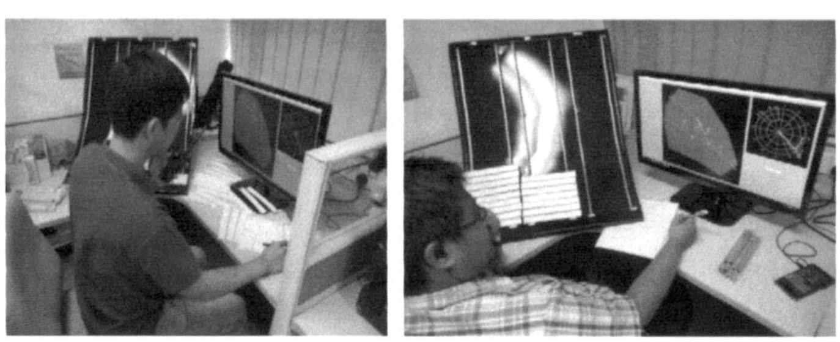

(a) (b)

Figure 10. Subject strategy during control: (a) Subject 1, (b) Subject 2

3.2. Trained subject

For the trained subject, based on the result as shown in Figures 11 and 12, it can be seen that Subject 3 (trained subject) recorded the highest number of conflict during the whole scenario. However, for low traffic, the subject was able to reduce the conflict to zero (Figure 12). Most of the conflicts were due to the intercepting route, which merges the aircraft into one route.

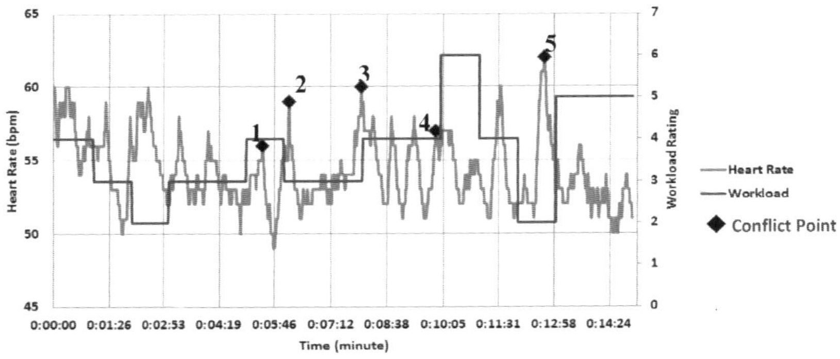

Figure 11. High traffic for Subject 3 using Procedural Control Bay

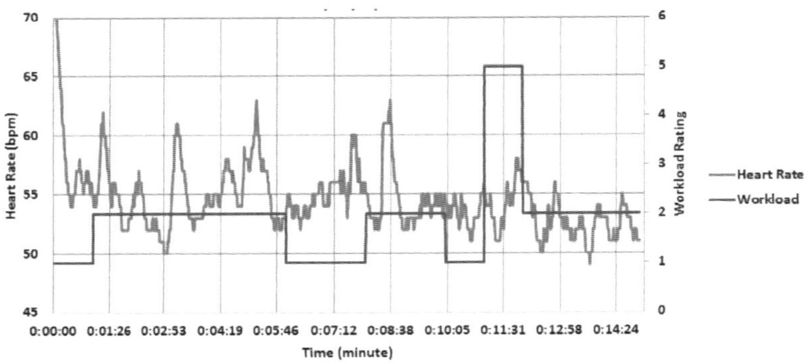

Figure 12. Low traffic for Subject 3 using Procedural Control Bay

The average rating of workload prompted by Subject 3 (trained subject) was higher than the Subject 1 (expert subject) for both high and low traffic. The most obvious difference between trained and expert subjects is the heart rate reading of trained subject increase during conflict occurrence regardless of their workload. This is due to the big gap of knowledge, experiences and skills of managing air traffic between the expert and trained subject. Two conflicts taken from the Subject 3 were shown in Figure 13.

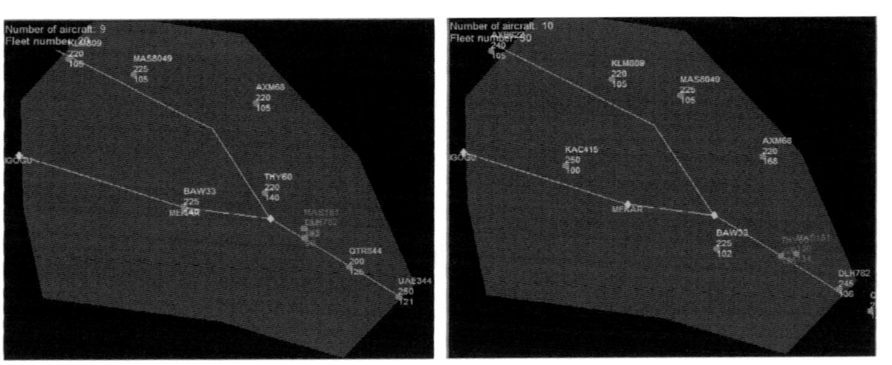

(a) Conflict 1 (b) Conflict 2

Figure 13. Conflicts from Subject 3 using Procedural Control Bay

For control activities using Electronic Flight Strip Bay, it is observed that Subject 3 has also resulted in five conflict situations as shown in Figure 14. The diamond point in the graph shows the time when the conflict occurred. It clearly shows that there is a positive association between workload rating and the heart rate of the trained subject at this particular time. The result in Figure 14 shows that the point of conflict is the same with the peak of the heart rate and the workload rating. Only one conflict point that resulted in no peak in the workload rating, which is conflict point 2. This is because the conflicting aircraft has already been handed over to other sector and the conflict was being totally neglected by the Subject 3 (Figure 15).

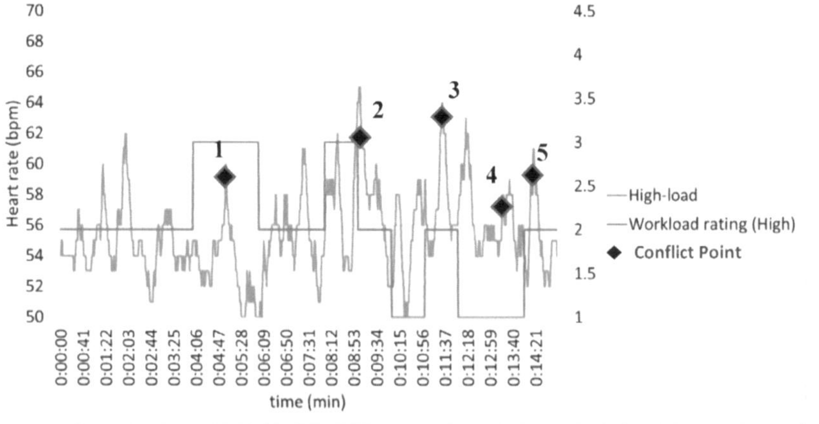

Figure 14. High traffic for Subject 3 using Electronic Flight Strip Bay

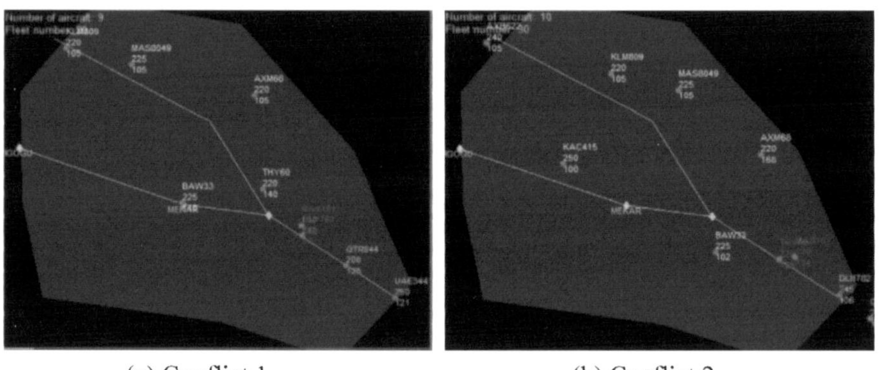

(a) Conflict 1 (b) Conflict 2

Figure 15. Conflicts from Subject 3 using Electronic Flight Strip Bay

3.3. Subjective workload measure (NASA TLX)

The NASA TLX score results after using the Procedural Control Bay show that the overall NASA TLX score for Subject 1 is 65.3, which was much higher than the overall NASA TLX score for Subject 2 (Figure 16). It was clear that Subject 1 faced difficulty or felt more pressured during the experiment than Subject 2. Subject 2, on the other hand, reported a low rating for mental, physical and temporal demand as in the NASA TLX score in Figure 16 (b). It is observed that Subject 2 have a more organized working method where he took out the strip from the bay once the aircraft was being handed over to the next sector, thus avoiding clutter on his Procedural Control Bay.

The same situation can be observed while using the Electronic Flight Strip Bay. Based on Figure 17, the overall NASA-TLX rating was significantly higher for Subject 1. His rating shows that the mental demand, physical demand and temporal demand was the highest rating compared to performance, effort

IOP Conf. Series: Materials Science and Engineering **405** (2018) 012004 doi:10.1088/1757-899X/405/1/012004

and frustration. Based on Figure 17, the overall NASA TLX score for Subject 1 is 64.7 whereas Subject 2 reported a NASA TLX score of 24.3. This is in accordance to the NASA TLX score given during the Procedural Control activities.

	Rating	Tally	Weight
Mental Demand	70	3	0.2
Physical Demand	70	3	0.2
Temporal Demand	70	3	0.2
Performance	35	2	0.13333
Effort	70	3	0.2
Frustration	70	1	0.06667
Overall = 65.33333			

(a)

	Rating	Tally	Weight
Mental Demand	30	5	0.33333
Physical Demand	20	3	0.2
Temporal Demand	30	2	0.13333
Performance	20	3	0.2
Effort	35	0	0
Frustration	30	2	0.13333
Overall = 26			

(b)

Figure 16. NASA TLX results using Procedural Control Bay for: (a) Subject 1, (b) Subject 2

	Rating	Tally	Weight
Mental Demand	60	3	0.2
Physical Demand	55	1	0.16667
Temporal Demand	60	2	0.13333
Performance	45	5	0.33333
Effort	60	4	0.26667
Frustration	55	0	0
Overall = 64.66666			

(a)

	Rating	Tally	Weight
Mental Demand	20	2	0.13333
Physical Demand	15	3	0.2
Temporal Demand	25	4	0.26667
Performance	15	2	0.13333
Effort	45	2	0.13333
Frustration	30	2	0.13333
Overall = 24.33333			

(b)

Figure 17. NASA TLX rating using Electronic Flight Strip Bay for: (a) Subject 1, (b) Subject 2

4. Conclusion and recommendation

This study was aimed on assessing the response behaviour of the use of the Procedural Control Bay as well as Electronic Flight Strip Bay with the assistance of radar information on controller through HITL experiment. The method of assessment for controller response behaviour was measured through heart rate readings, workload rating and also questionnaires from NASA-TLX. The HITL experiment were subjected to two experts and two trained subjects through scenarios of high and low traffic feeds. Based on the results, high traffic scenario was observed to result in a higher heart rate peak together with a higher workload rating. Meanwhile, for the low traffic scenario, it was observed that a lower workload rating was consistently prompted by the subjects along with a more stable heart rate readings. The results were observed to be consistent for both Procedural Control Bay as well as Electronic Flight Strip Bay. The promising outcome of this initial study suggests that heart rate monitoring device may be suitable in workload monitoring of ATCo. However, a more comprehensive study needs to be conducted with a greater number of participants, especially from the expert subjects group. It is also recommended that the HITL experiment is conducted on-site as to properly investigate the suitability of the device to be used in real condition.

References

[1] Hofmann T, König C, Bruder R, Bergner J. How to reduce workload - Augmented reality to ease the work of air traffic controllers. Work 2012; 41(SUPPL.1): 1168–1173

[2] Suárez S, López N, Puntero P, Rodriguez E. Quantifying air traffic controller mental workload. 4th SESAR Innov. Days, 2014

[3] Wang Y, Zhang Q, Zhu C, Hu M, Duong V. Human activity under high pressure: A case study on fluctuation scaling of air traffic controller's communication behaviors. Phys. A Stat. Mech.

its Appl. 2016; 441: 151–157
[4] Jou R, Kuo C, Tang M. A study of job stress and turnover tendency among air traffic controllers: The mediating effects of job satisfaction. Transp. Res. Part E Logist. Transp. Rev. 2013; 57: 95–104
[5] Malakis S, Kontogiannis T. A sensemaking perspective on framing the mental picture of air traffic controllers. Appl. Ergon. 2013; 44(2): 327–339
[6] Li Q, Zhang Y, Su R. A flow-based flight scheduler for en-route air traffic management. IFAC-PapersOnLine 2016; 49(3): 353–358
[7] Truitt T, Albright C A, Barile A B, Vortac O U, Manning C A. How controllers compensate for the lack of flight progress strips, 1995
[8] Duca G, Attaianese E. The realistic consideration of human factors in model based simulation tools for the air traffic control domain. Work 2012; 41(Suppl 1): 145–150
[9] Majumdar A, Ochieng W, Bentham J, Richards M. En-route sector capacity estimation methodologies: an international survey. Journal of Air Transport Management 2005; 11(6): 375–387
[10] Ahlstrom U. An eye for the air traffic controller workload. J. Transp. Res. Forum 2007; 46(3): 103–108
[11] Djokic J, Lorenz B, Fricke H. Air traffic control complexity as workload driver. Transp. Res. Part C Emerging Technology 2010; 18(6): 930–936
[12] Lee Y S, Pathirana P N, Caelli T. Doppler radar in respiratory monitoring: detection and analysis, 2013.
[13] Lee P, Smith N, Bienert N, Brasil C, Buckley N, Chevalley E, Homola J, Omar F, Parke B, Yoo H. A human-in-the-loop evaluation of a coordinated arrival / departure scheduling operations for managing departure delays at LaGuardia Airport, 2016
[14] https://ext.eurocontrol.int/ehp/?q=node/1585
[15] Gawron V J. Human performance measures handbook, 2000

IOP Conf. Series: Materials Science and Engineering **405** (2018) 012005 doi:10.1088/1757-899X/405/1/012005

Study the effect of cooling medium on the torsion resistance and hardness of medium carbon steel

Ahmed M. Abaas, Ayad A. Ramadhan*, Fattah H. Hasan

Department of Mechanical Engineering, Al-Hawija Technical Institute, Northern Technical University, Iraq

*ayad_almuhndis@yahoo.com

Abstract: In this paper, the effect of the cooling medium on the strength of torsion and hardness is discussed. The specimen used in this study is of medium carbon steel grade. The carbon percentage was determined as 0.55% C through the Arl spectroscopy instrument. The specimen dimension is 6 mm in diameter, gauge length is 76 mm and the total length is 144 mm. Different cooling mediums such as water, air, furnace atmosphere and oil were used to determine the effect of cooling speed on mechanical properties, specifically the torsion strength and hardness. In all treatments, the holding period was 30 minutes. From the results, cooling in the furnace atmosphere had the greatest impact on the resistance of the torsion of the carbon steel from other cooling methods. The non-heat-treated sample was the least resistant between the samples whereas the water-cooled sample had the highest in terms of their scratch resistance according to the results of the hardness device.

1. Introduction

Some researchers had investigate the effects of heat treatment on the mechanical properties of the medium carbon steel [1-3]. For instances, samples of medium carbon steel were tested after heated at 9000°C and soaked for 60 minutes in a muffle furnace and quenched in oil, while other researchers studied the effect of tempering and austenizing on the tensile strength and the toughness of the fracture on the 300M steel. The results showed that an appropriate heat treatment could be used to improve toughness and tensile strength properties in the same process. There was 20 % of the practical examination leads to the non-loss of tensile property by increasing austenizing temperature from 871 to 982 and more [4]. In addition, some researchers investigated a practical work to study the effect of hardness and roughness on the torsion strength of low carbon steel (1020). The results showed that ductility decreases by increasing the roughness and hardness of the sample surface. On the other hand, shear strength increases with increasing surface hardness. The strength is found to increase with increasing hardness and roughness, and the reason for this is because of the high concentration of stresses due to the sharp surface shape [5].

The effect of the cooling mediums such as water, furnace and air on mechanical properties of different carbon steels was investigated. The steel was heated to 977°C for four hours and the results showed that there was a clear difference in the microstructure of different carbon steels (0.2, 0.4 and 0.6% C) with the cooling medium. The hardness of the water-cooled specimen was increased for all various carbon steel [6]. Fadare et al. (2011) studied the effect of heat treatment on steel (37-2 NST) [7]. Different cooling mediums were used: water, air and furnace atmosphere. It was observed that there was increased hardness and yield, as well as tensile strength, with increased plastic deformation. On the other hand, the impact was reduced due to the effect of hardening. The normalizing processes increase the tensile strength as well as hardness more in comparison to annealing processes. Relationship between tensile strength and hardness in sheet of steel containing martensite and ferrite phases was derived. The relationship could be used as an approximate value when applying a Vickers test on the samples, in which the accuracy of the results was approximately 10%. Moreover, researchers had also found a method to calculate yield and strain depending on Vickers hardness results and converted them to the Brinell hardness for more benefits to build a diagram between hardness and tensile strength for some metals [8]. Furthermore, some researchers also studied the effect of cooling medium (polymer polyethylene with change of concentration) on the mechanical properties (i.e. impact, tensile strength and hardness-HB) of low carbon steel [9]. The samples were quenched to 575°C and also 220°C for a half an hour, and it was found that polymer quenching has an excellent effect on the mechanical properties as a slow cooling. By increasing the polymer concentration, mechanical properties can be increased [9].

The effect of different types of heat treatments on mechanical properties such as Rockwell hardness and impact was also investigated. The cooling mediums used were water and air. The treatment of tempering and quenching were used in this research apart from hardening. SEM was used in the microstructure. The researchers found that the highest value of the hardness was from those water-cooled samples whereas the impact from those treated in a tempering method was the best [10]. Investigation on the effect of vanadium (0.28 and 0.15) % as an alloying element in medium carbon steel with different cooling mediums to test a mechanical property (hardness and microstructures) was conducted and the results showed that increased vanadium in steel caused an increasing of forms ferrite like acicular with low cooling. When cooling rate was above the critical temperature, the hardness also increased but with high cooling, vanadium had greater impact on the hardness [11]. A useful study by Verma and Kumar (2013) on medium carbon steel with 0.4% carbon to achieve properties like hardness, elongation and tensile test by a suitable heat treatment (tempering) about 900°C and cooled by water and then tempered on multi temp degrees. The results showed a decreasing value of hardness with high temperature of tempering, elongation would be increased with high temperature of tempering too. By increasing the time and temperature of tempering, strength of tensile have been lower value [12]. The tensile resistance under the influence of high heating and wear abrasive resistance of the medium carbon steel with tiny alloys was discussed in the published work of Gündüz and Acarer (2006). Heat treatment was used on specimens that were treated and without treatment. The researcher concluded that medium carbon tiny alloys have dynamic resistance of ageing strain due to the effect of dislocation and bonding atoms. Wear abrasive had also increased by heating up to 300°C [13]. In the meantime, mechanical properties like ductility and impact of the medium carbon steel bars under the influence of the heat treatment were discussed by Daramola et al. (2010). In this study, the samples were heated to 830°C until they reached the austenite zone. Using water as the quenching medium, the samples were heated again at 480°C. In this study a good impact resistance was found as well as ductility [14].

An investigation on heat treatments of 12CrMoV steel on its mechanical properties and also behavior of fracture was made whereby a sequence of heat treatment produced different results. Cooling mediums such as water, air and oil had an influence on some mechanical properties. It was noted that the impact been in a good state by oil quenching from 1050°C and then tempering on 700°C, while cooling by air from same temperatures produced a low impact resistance. An alpha phase was produced as a precipitate by heating temperature up to 1200°C [15]. Meanwhile, a carbon steel with less than 0.1% C was investigated by another study that focused on the effect of heat treatments on mechanical properties. After heating the samples, a tempering was applied in different temperatures with different cooling time. Strength and yield were lowest when the temperature of tempering was increased but the ductility property was enhanced [16]. Heating or cooling of a metal can change its micro structure, which causes variations in the mechanical and physical properties and affects the behavior of the metal in processing and operation. By using the heat treatment, metal properties can be improved for the application in engineering fields. There are different spare parts used for several different purposes collected from foreign countries in Spinning Mills available machine in Bangladesh. Spindle is an important part of spinning mills [17].

Previous research has examined mechanical properties such as tensile and bending and how they are affected after the heat treatment as well as cooling type. This current research is characterized by the study of resistance to torsion and how to improve the different cooling types in addition to hardness. The aim of this study is to investigate the effect of cooling medium on some mechanical properties such as resistance to torsion and hardness of the medium carbon steel. The chemical test was carried out in advance and the dimensions of the samples used were: d = 6 mm, gauge length = 76 mm, total length = 144 mm. Multiple cooling modes (water, air, oil, and cooling inside the oven) were used in this research. Also, the number of cycles in the torsion test was calculated until the samples failed. The devices used in this study are torsion tester, electric oven and hardness test.

2. Theoretical calculations

2.1. Study of torsion

The forces in this test are inverse and parallel, and do not fall on the longitudinal axis of the body. Thus, a double beam will be generated by the twist on the longitudinal axis. The amount of the torque that is causing the torsion is equal to that of one of the opposing forces in the distance between them. If two equal forces are placed at the end of a rod and the other end is fixed, the determination of the duality arising from these two forces, which is equal to the sum of one of the two forces in the length of the distance between them and the effect of twisting the penis around its longitudinal axis. The value of torque and output of the effective weights is calculated from the following Eqn. 1, R = radius of pulley, m = weights and g = constant of acceleration. Furthermore, the shear stress and shear strain values are calculated by Eqn. 2 and Eqn. 3, respectively, with the twist of angle, $\emptyset$ is given in Eqn. 4. Last but not least, the value of modulus of rigidity is given by Eqn. 5.

$$T = m\,g\,R \tag{1}$$

$$\tau = \frac{16\,T}{\pi d^2} \tag{2}$$

$$\gamma = \frac{r\emptyset}{L} \tag{3}$$

$$\emptyset = N\,d\,\frac{\pi}{180} \tag{4}$$

$$G = \frac{\tau}{\gamma} \tag{5}$$

2.2. Study of hardness (Brinell)

One of the methods or tests used to measure the hardness of a material is invented by the Swedish engineer John August Brinell, which is the test method of compressing a ball of hard steel or a ball made of tungsten carbide material on flat sample of the substance to be tested. In this test, the thickness of the sample shall be at least 10 times the depth of the trace (the mark left by the ball in the sample). After the diameter of the trace is set for the test, the hardness is calculated by using Eqn. 6 where *BHN* is the Brinell hardness number, *P* is the load in kg, *D* is the diameter of the ball used as a harder in millimeters and *d* is the diameter of the trace on the surface of the sample in millimeters.

$$BHN = \frac{2P}{\pi D\left(D - \sqrt{(D^2 - d^2)}\right)} \tag{6}$$

3. Experimental methodology

The aim of the experimental study that is presented in this work is to test the heat treatment. The temperature of the metal or alloy was raised to a certain temperature (850°C). After that, the temperature was maintained for an appropriate period of time. Next, the metal or alloy was cooled at a certain cooling rate. These three factors: the temperature at which the metal is heated, temperature holding time and cooling rate; have a significant impact on the properties to be achieved from the heat treatment. Therefore, these factors should be specified accurately. The hypo eutectoid steels are heated by 30°C above the upper critical temp (A3). The temperature is chosen according to the ratio of the carbon such that the ferrite phase is converted to an austenite.

3.1. Types of cooling medium
The cooling medium should have a relative advantage of absorbing the amount of heat in the first moments of the cooling process to ensure the Martensite phase. Then, a low cooling rate is necessary to prevent any deformation that may occur in the working area. The cooling media used were water, oil, salt, molten salt. Carbon steel with a carbon content of less than 0.25% is not acceptable for heat treatment. The higher the temperature of the cooling medium, the lower the cooling rate. To ensure this, large quantities of cooling medium should be used and the creation of spiral movements during the flipping of work pieces.

3.2. Metal selection
In this study, samples of medium carbon steel were used. Preparation of the steel samples (i.e. bar cutting and manufacturing process) was in accordance with ASTM standard specifications. The samples were in the following dimensions: diameter (6 mm), gauge length (76 mm), final length (144 mm); arranged for use in heat treatment. In general, as shown in Figure 1, the sample was a solid shaft with its end having a diameter greater than the center to avoid fracture at the end of the sample. After the chemical examination of the sample by using Arl spectrometer, the details obtained are presented in Table 1.

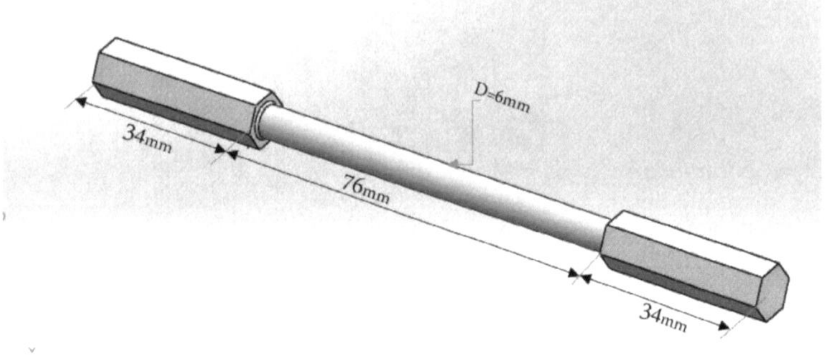

Figure 1. Specimen dimensions

Table 1. Chemical composition of medium carbon steel specimen

Element	C	Mn	Cr	Mo	Ni	Si	P
Weight fraction (W_t)	0.55%	0.81%	1.8%	1.04%	0.68%	0.21%	-------

The samples were cut into the specifications and required dimensions using the lathe machine as shown in Figure 2. A cylindrical rod is used with the lathe machine and with a good surface finish.

3.3. Heat treatment
The principle of full annealing includes heating to the austenite phase depending on the carbon percentage, removing all the residual stresses in steel, heating the samples above the upper critical temp (A3). The high temperature required in this work was approximately 850°C. The electric furnace was used to complete the

required heat treatments as shown in Figure 3. The specimen under testing by torsion and Brinell hardness device is shown in Figure 4 and Figure 5, respectively.

Figure 2. Specimen machining and finishing by turning device (lathe)

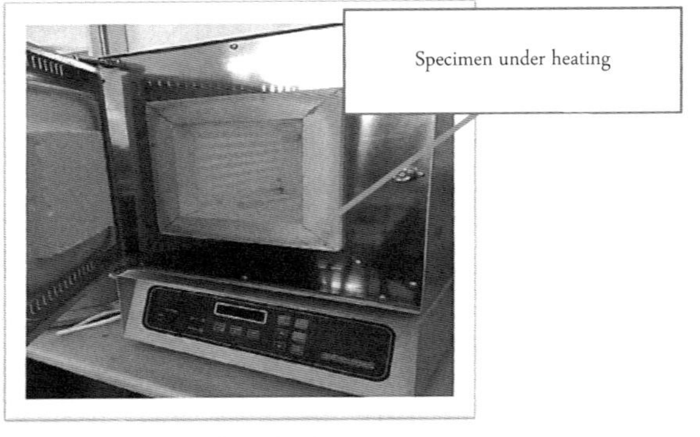

Figure 3. The furnace used in the heat treatment process

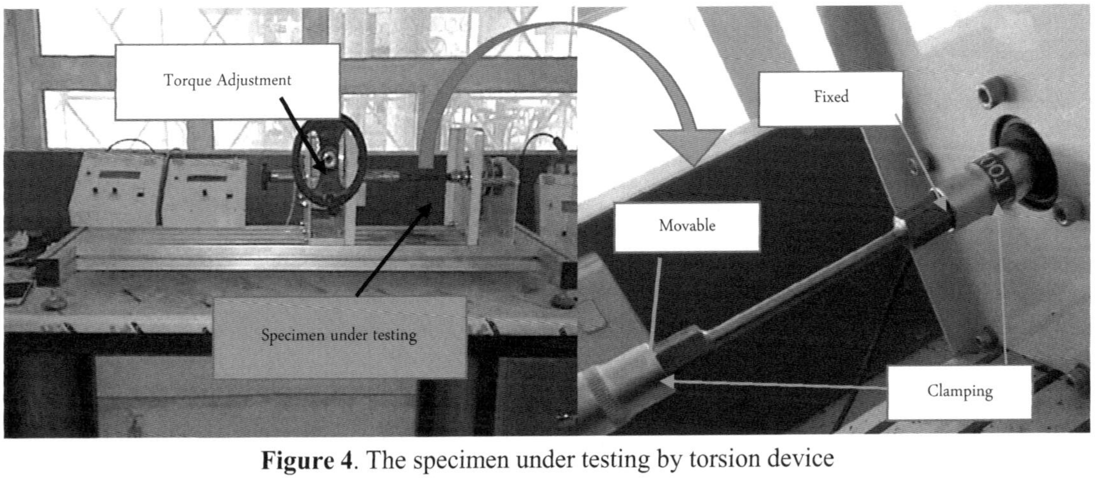

Figure 4. The specimen under testing by torsion device

Figure 5. The Brinell hardness test

4. Results and Discussion

Through the results of the effect of torque with the twist angle (rotation angle) as well as the effect of the type of cooling medium, the following observations were made. As shown in Figure 6, the water-cooled sample has the highest torsion values compared to the other samples, reaching the highest value (55.9) with the angle of torsion. This is due to the ability of the alloy to withstand the rotational momentum due to the formation of the martensite phase, which is transformed from austenite and ferrite phases, which is formed by the high rate, which in turn prevents the rapid spread of atoms to form the fe3c cement. The oil-cooled sample will be the least thereafter because of the lower cooling rate if higher hardness can be obtained as the cooling is faster. The samples that have not been subjected to heat treatment are the lowest among the other samples and the heat treatment is up to (28.3). This in turn reflects the importance of our research in obtaining higher torsion resistance.

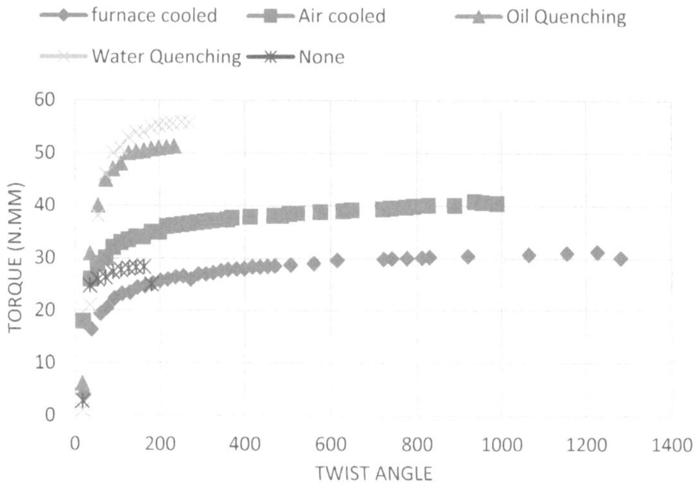

Figure 6. The relationship between torque and twist angle for different cooling mediums

Furthermore, from Figure 7, the furnace cooling has maximum stress resistance proportional to the shear strength until failure due to the fine pearlite due to the slow cooling of the furnace. The sudden water cooling is of higher value for cut stress and early failure, with shear behavior and rapid failure is due to the fact that the behavior of the water-cooled sample is like that of fragile materials with the presence of the Martensite. The cooled samples of oil and air subsequently follow in terms of the least shear stress value, while the non-treated heat sample is the least of the rest.

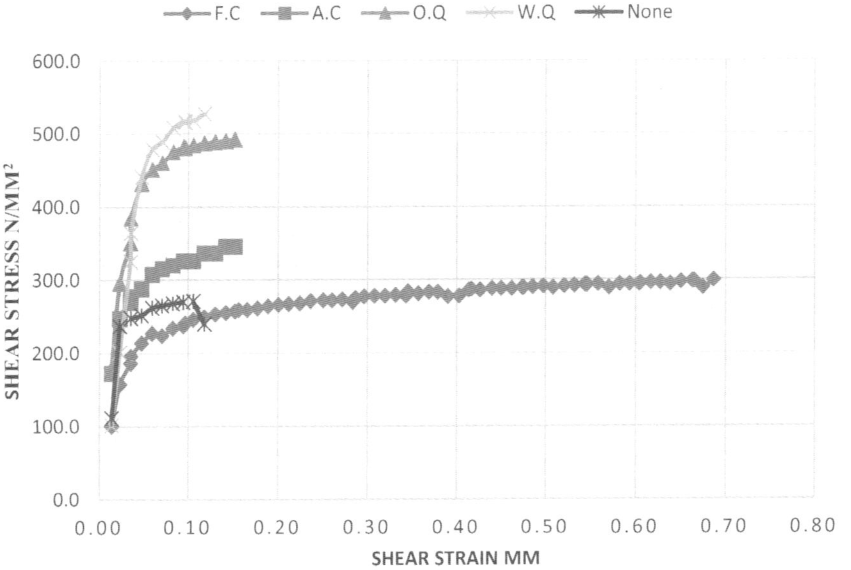

Figure 7. Relation between shear stress and shear strain for different cooling mediums

Meanwhile, Figure 8 shows that the rotational torque resistance of the five samples is different since the chilled sample of the furnace has higher rotational torque resistance until the fracture occurs, thus increasing the number of cycles up to 213 cycles. This is due to the fact that slow cooling is successive phase of ferrite and cementite, combining the softness of the ferret and the hardness of the cementite and is in the form of a soft perlite phase. Air cooling is second best with the number of cycles 171 while the sample cooled with oil and non-heat treatment are the lowest. Table 2 shows the hardness of the samples with the water-cooled sample has a high hardness up to 1401 with the hardness of Brinell. The presence of the martensite phase, which is formed by rapid cooling gives the hardness and resistance to scratching.

Based on the readings of the device during the rotation of the lever, the reading of the torque and the torsion angle were taken during a complete cycle with the torque, which is equivalent to 1/6 cycles in the same sample. Table 3 tabulates the type of cooling methods used in the practical part after heating to the required crystallization temperature of 850°C, depending on "the carbon content of the sample which is approximately 0.55". The duration of the fixation (installation of the sample in temperature 850°C) is about half an hour depending on the thickness of the sample to ensure the occurrence of adequate crystallization process.

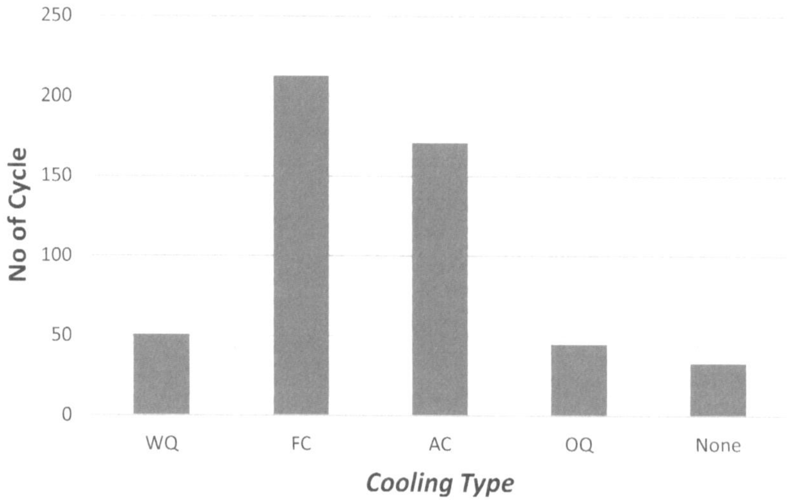

Figure 8. The relation between numbers of cycles and cooling type until specimens fail

Table 2. Tests results of some mechanical properties for tested samples

Conditions	HB (Kg. F/mm2)
Furnace cooled	205
Air cooled	456
Oil quenched	658
Water quenched	1401
As received	175

Table 3. Classification of samples

The samples symbol	Samples condition
FC	heated to (860°C), for a 30 minutes and furnace cooled
AC	heated to (860°C), for a 30 minutes and air cooled
WQ	heated to (860°C), for a 30 minutes and water quenched
OQ	heated to (860°C), for a 30 minutes and oil quenched
None	Samples without heat treatment (as received)

5. Conclusions and recommendations

The results from this research have demonstrated that the torsion resistance increases as the cooling rate increases, reaching its highest value when cooling with water. The test device used continues to measure the angle of torsion with a low cooling rate until the break occurs as with the cooling oven atmosphere. It is observed that the value of shear stress is the highest value with rapid cooling and the rate of hardness increases with increasing cooling rate.

For future study, it is recommended to use other methods of test such as fatigue and impact. In addition, other types of cooling mediums such as salty water and cold water can be used, as well as sand. Other types of heat treatments can also be used such as surface hardening.

Acknowledgements

The authors wish to thank North Gas Company laboratories of the Iraqi Ministry of Oil for providing the full requirements of the experimental work in this research located in the laboratories of the Faculty of Engineering / University of Kirkuk

References

[1] Senthikumar T, Ajiboye T K. Effect of heat treatment processes on the mechanical properties of medium carbon steel. Eff. Heat Treat. Process. Mech. Prop. Mediu. Carbon Steel 2012; 11(2): 143–152

[2] Nayak S S, Anumolu R, Misra R D K, Kim K H, Lee D L. Microstructure-hardness relationship in quenched and partitioned medium-carbon and high-carbon steels containing silicon. Mater. Sci. Eng. A 2008; 498(1–2): 442–456

[3] Murugan V K, Koshy Mathews P. Effect of tempering behavior on heat treated medium carbon (C 35 Mn 75) steel. Int. J. Innov. Res. Sci. Eng. Technol. 2013; 2(4): 945–950

[4] Youngblood J L, Raghavan M R. Influence of heat treatment on mechanical properties of 300M steel. 16th Structural Dynamics and Materials Conference, 1975

[5] Zurita Hurtado O J, Di Graci Tiralongo V C, Capace Aguirre M C. Effect of surface hardness and roughness produced by turning on the torsion mechanical properties of annealed AISI 1020 steel. Rev. Fac. Ing. Univ. Antioquia 2017; 84: 55–59

[6] Çalik A. Effect of cooling rate on hardness and microstructure of AISI 1020, AISI 1040 and AISI 1060 Steels. Int. J. Phys. Sci. 2009; 4(9): 514–518

[7] Fadare D A, Fadara T G, Akanbi O Y. Effect of heat treatment on mechanical properties and microstructure of NST 37-2 steel. J. Miner. Mater. Charact. Eng. 2011; 10(3): 299–308

[8] Gasko M, Rosenberg G. Correlation between hardness and tensile properties in ultra-high strength dual phase steels–short communication. Mater. Eng. 2011, 18: 155–159

[9] Chandan B R, Ramesha C M. Evaluation of mechanical properties of medium carbon low alloy forged steels quenched in water, oil and polymer. J. Mater. Sci. Eng. 2017; 6(2): 1–5

[10] Ismail N M, Khatif N A A, Kecik M A K A, Shaharudin M A H. The effect of heat treatment on the hardness and impact properties of medium carbon steel. IOP Conf. Ser. Mater. Sci. Eng. 2016; 114(1)

[11] Hui W, Zhang Y, Shao C, Chen S, Zhao X, Dong H. Effect of cooling rate and vanadium content on the microstructure and hardness of medium carbon forging steel. J. Mater. Sci. Technol. 2016; 32(6): 545–551

[12] Verma A, Kumar P. Influence of heat treatment on mechanical properties of Aisi1040 steel. IOSR Journal of Mechanical and Civil Engineering 2013; 10(2): 32–38

[13] Gündüz S, Acarer M. The effect of heat treatment on high temperature mechanical properties of microalloyed medium carbon steel. Mater. Des. 2006; 27(10): 1076–1085

[14] Daramola O, Adewuyi B, Oladele I. Effects of heat treatment on the mechanical properties of rolled medium carbon steel. Journal of Minerals and Materials Characterization and Engineering 2010; 9(8): 693-708

[15] Bashu S A, Singh K, Rawat M S. Effect of heat treatment on mechanical properties and fracture behaviour of a 12CrMoV steel. Mater. Sci. Eng. A 1990; 127(1): 7–15

[16] Kang S S, Bolouri A, Kang C. The effect of heat treatment on the mechanical properties of a low carbon steel (0.1%) for offshore structural application. Proc. Inst. Mech. Eng. Part L J. Mater. Des. Appl. 2012; 226(3): 242–251

[17] Kowser M A, Motalleb M A. Effect of quenching medium on hardness of carburized low carbon steel for manufacturing of spindle used in spinning mill. Procedia Engineering 2015; 105: 814–820

IOP Conf. Series: Materials Science and Engineering **405** (2018) 012006 doi:10.1088/1757-899X/405/1/012006

Statistical evaluation of damage size based on amplitude mapping of damage-induced ultrasonic wavefield

GAN Chia Sheng [1], CHIA Chen Ciang [1,*], TAN Lin Yaw [1], Norkhairunnisa MAZLAN [1], Joel B. HARLEY [2]

[1] Dept. of Aerospace Engineering, Universiti Putra Malaysia, Selangor, Malaysia
[2] Dept. of Electrical and Computer Engineering, University of Florida, Florida, USA

* chia_cc@upm.edu.my

Abstract. The inability to statistically evaluate the damage size based on amplitude mapping has long plagued the ultrasound propagation imaging (UPI) system and related non-destructive evaluation (NDE) community. This paper proposes a damage visualization method called the Statistically Thresholded Anomaly Mapping (STAM) method to solve this problem. It could isolate the damage-induced anomalous waves from an ultrasonic full wavefield data, map their amplitudes into an image and statistically differentiate damage from the background noise and pristine region of a specimen, without prior knowledge or reference data of the specimen being inspected. The proposed method was demonstrated through the visualization and evaluation of a thermal damage inflicted in a glass fibre reinforced composite specimen. The effects of two parameters on result quality and accuracy of damage size evaluation were studied by varying their values independently. The optimized result showed that 99.9% of the background noises could be removed while maintaining clear visualization of damage, hence allowing the users to evaluate the presence, location, shape, size and severity of the damage accurately.

1. Introduction

Ultrasounds are widely used in the material and structural nondestructive evaluation (NDE) due to its sensitivity towards physical properties such as volume, thickness, temperature and most importantly, damages or anomalies that are present on the surface or within the material [1]. In order to harvest these information that ultrasounds carry, the ultrasound propagation imaging (UPI) system had been developed [2] for NDE of plate-like waveguides such as aircraft wings [3] and radome panels [4]. UPI system and its associated result processing methods had shown the capabilities of revealing damages such as small cracks through the detection of anomalous waves, which corresponds to the scattering and reflected waves when an ultrasonic wave encounters a damage [5,6]. In recent years, UPI system had been further improved to give higher visibility to these anomalous waves. The result processing methods, which suppress incident waves through the subtraction of adjacent waves [5], frequency-wavenumber domain filtering [7,8] and model-driven wavefield baseline subtraction [9], proved that the suppression of incident waves reveals anomalous waves clearly, thus revealing the presence and location of damage.

However, these methods do not have statistical reliability when it comes to damage sizing and the results are typically in the video forms where the inspectors have to estimate the size of a damage by repeatedly examining the result videos, which is both unreliable and also inaccurate. This becomes a significant problem when the damage is incorrectly evaluated and its severity is underestimated. For example, aircrafts are likely to get struck by lightning once every 10,000 hours of flight on average [10], which for an aircraft that is built mostly of composites such as Boeing 787, it is highly hazardous to underestimate the inevitable lightning damage as the lightning damages can severely degrade the mechanical properties of composites [11, 12]. Tragedies involving the loss of countless lives might occur if the composites' damages deteriorate its mechanical strength to a critical level and causes a failure mid-flight. Therefore, it is of utmost importance to have statistical reliability when it comes to the detection and evaluation of damages.

Previously, Capriotti et al. (2017) used Multivariate Outlier Analysis (MOA) to statistically analyse ultrasonic signals in comparison to an ultrasonic signal obtained from a pristine region of the same specimen [13]. Through the use of the Mahalanobis Squared Distance as part of the MOA, a Damage Index (DI) was obtained, whereby a large value of DI represents a damage. In their paper, MOA is reported to have 100% probability of detection while having a 10% probably of false alarms. Although this method has a high probability of detection, there exists an uncertainty of whether the pristine region of specimen used to obtain the reference signal is truly free from manufacturing defects. If such pristine region indeed exists, then paradoxically, its location must be known prior to the processing of inspection data. Other than that, the outcome of the analysis is unable to determine the sizes of the damages, which is a crucial information in order to evaluate the severity of any damage.

To solve the inability of obtaining a statistically reliable result, this paper describes a reference-free method that provides a statistical damage threshold to the results obtained from amplitude mapping of the damage-induced ultrasonic wavefield, where values higher than the threshold are considered as a damage. Through the use of the threshold, the amplitude map will be rid of noises, leaving an amplitude map that shows only the damages. This method allows the damage sizes to have an absolute value instead of undesirable and indecisive estimations by examining the result videos repeatedly.

2. The proposed method

The proposed method is called Statistically Thresholded Anomaly Mapping (STAM). It is suitable for processing of any ultrasonic full wavefield data acquired using UPI system, regardless if the ultrasonic excitation-sensing is achieved using a scanning laser exciter and a fixed piezoelectric sensor, or a fixed piezoelectric exciter and a scanning laser Doppler vibrometer (LDV) sensor. This method has an anomalous waves isolation component and a damage threshold calculation component, described in the following sections.

2.1. Anomalous waves isolation

Due to the dispersive nature of ultrasound, waves guided by plate-like waveguides often have multiple modes mixed or overlapped in time [14, 15]. Wave modes which exist due to inspection excitation are collectively known as the incident modes. When these incident modes reflected by the damages have insufficient propagation time to separate from each other near damages, mode conversion occurs [16]. These mode-converted waves, also known as the anomalous waves, typically have a magnitude of a few orders smaller than the incident ones. Their small magnitude and the mix-modal characteristics make them unnoticeable at all or hard to be noticed, hence making the detection of damage associated with these anomalous waves difficult.

In order to isolate the anomalous waves, the acquired ultrasonic wavefield signals are first grouped together as a three-dimensional (3D) spatiotemporal data matrix. After that, the 3D data undergoes a 3D DFT (discrete Fourier transform) to transform from the space-time domain to the wavenumber-frequency domain [17–19]. Once transformed, the incident modes are usually demixed and easily distinguishable in the wavenumber-frequency domain. They are then suppressed or filtered out using a mode filter described in [19]. For thin plate-like materials typically used in the aerospace structural applications, the maximum number of modes exists is usually two, hence only one step of filtering is needed. As long as the wavenumber and the frequency of the anomalous waves are different from the incident modes, the anomalous waves will remain in the mode-filtered data, hence isolated for further processing.

The process of rendering the anomalous waves into an amplitude map to show the presence and location of damages can be done by determining the peak-to-peak amplitudes of all mode-filtered signals and mapping the values into an intensity graph while conserving their relative spatial position. By using suitable colour scale, a stark contrast between the damage and pristine areas can be realized, which allows one to easily identify the damaged areas from the pristine areas. This is because after the removal of incident modes, only the anomalous waves and noises remain in the 3D data matrix, with damaged areas containing the anomalous waves and the pristine areas containing the noises. These anomalous waves' amplitudes are higher than the noise floor. When a suitable colour scale is used, the damages can be easily identified. A suitable variable time window, as described in [5], such as where the peak-to-peak amplitude of anomalous waves are at the highest and the noise at the lowest, can also be used optionally to increase the contrast between the peak-to-peak amplitudes of anomalous waves and noises so that the clearest result can be produced. This anomalous waves isolation method offers the advantage of not being dependent on correct selection of any wave mode for damage evaluation, hence circumvented the need of performing reference inspection of the specimen at pristine state to understand the complex ultrasonic waves and modes.

2.2. *Damage threshold calculation*

To distinguish damages from background noises in an amplitude map, a statistical damage threshold is proposed. The damage threshold (*DT*) is defined as in Eqn. 1, where μ and σ represents the mean and standard deviation of background noises, respectively, while M represents a constant multiplier.

$$DT = \mu + M\sigma, \tag{1}$$

To determine μ, each pixel value is first ranked from the lowest value to the highest value. After that, the sum of values starting from the lowest rank up to the Nth rank is divided by N. The amplitude map is then masked according to Eqn. 2, where $V(x,y)$ represents the value of a pixel at location (x,y).

$$V(x,y) = \begin{cases} 0, & V(x,y) \le DT \\ V(x,y), & V(x,y) > DT \end{cases}, \tag{2}$$

The final output is an amplitude map devoid of noises such that the damages can be visualized, and their sizes and also locations can be accurately determined. The damage threshold is suitable for any anomalous waves amplitude map as long as the N number of pixels do not contain anomalous wave values.

3. Experimental validation

3.1. *Acquisition of ultrasonic wavefield data*

An ultrasonic full wavefield data matrix was acquired using an Ultrasound Propagation Imaging (UPI) system [2]. The system consists of a custom-made diode-pumped solid-state Nd:YAG (neodymium-doped yttrium aluminium garnet) pulsed laser (PL) as the source for ultrasonic waves on the specimen, a galvano-motorized laser mirror scanner (LMS) to point the laser to the intended impingement grid points, a fixed broadband ultrasound sensor with 350kHz cut-off to detect ultrasonic waves originating from the impingement points, an oscilloscope to digitize the detected waves from the sensor, and a computer for hardware synchronization, data storage, result processing, and result display. The PL has a wavelength of 532nm, a beam diameter of 4mm, a divergence of 0.5mrad and a pulse duration of 30ns, which is capable of producing ultrasonic waves non-destructively at the impingement point due to the transient thermoelastic expansion and contraction phenomenon [3]. The PL fires the laser at the LMS that is specially made to reflect lasers with 532nm wavelengths. The two computer-controlled galvano-motorized mirrors inside the LMS, which stand orthogonally from each other, directs the laser to the intended impingement grid points by rotating the mirrors they are attached to. By controlling the LMS, the laser impinges the area of inspection interest in a raster pattern, starting from the top left corner to the bottom right corner in an x by y grid with the same scan distance intervals for both the x and y axes.

From each laser impingement point, ultrasonic waves propagate and disperse into mixed multimodal ultrasonic waves that are detected by the sensor. The ultrasonic signals are then received by the oscilloscope to be digitized and sent to the computer for storage and processing.

3.2. Specimen

Specimen used in this study was fabricated in-house from a 250 by 250mm glass fibre reinforced epoxy prepreg (HEXCEL, BMS 8-79 Style 1581 with HexPly F155 resin system) as a 1-layer [(0/90)] lamina. Curing was done in vacuum bagging at 74 kPa, heated by thermal blanket at a rate of 4°C/min, followed by 90 minutes dwell at 127°C, and cool down to room temperature at 3°C/min. The thermal damage of the specimen was inflicted by contact heating using a 20mm diameter aluminium rod heated to 1.5 times of Tg (glass transition temperature), i.e. 182°C, for 30 minutes. Inspection scan was conducted on the specimen over an 80 by 80mm area with an interval of 0.5mm in both the x and y axes. Based on origin located at the lower left corner of inspection area, the center of the thermal damage was located at position (47, 35) mm and the sensor located at position (45, -105) mm. The acquired ultrasonic signals were then processed using the Statistically Thresholded Anomaly Mapping (STAM) method.

4. Results and discussion

4.1. Effects of anomalous waves isolation

Figure 1 shows a representation of frequency-wavenumber data in grey-scale, where signals with the highest and lowest amplitude are shown in white and black, respectively. Figure 1(a) clearly shows the incident wave mode as white lines spanning from $0m^{-1}$ to $200m^{-1}$ in the wavenumber (K) axis and 0kHz to 350kHz in the frequency (F) axis. It should be noted that the abrupt stop at 350kHz was caused by the sensor's detection range of up to 350kHz. Figure 1(b) shows the frequency-wavenumber (FK) map after filtering out the incident mode through the use of a wavenumber filter with a $37.5m^{-1}$ half-power taper bandwidth (B_M) and $12.5m^{-1}$ upper (C_H) and lower (C_L) wavenumber cutoffs, along the white FK lines that represent the incident mode. The filter was designed to radiate from the origin of FK axis to the infinity, suppressing all signals within its effective band to low amplitude, hence the filtered zone appeared as black regions extending to the top of Figure 1(b).

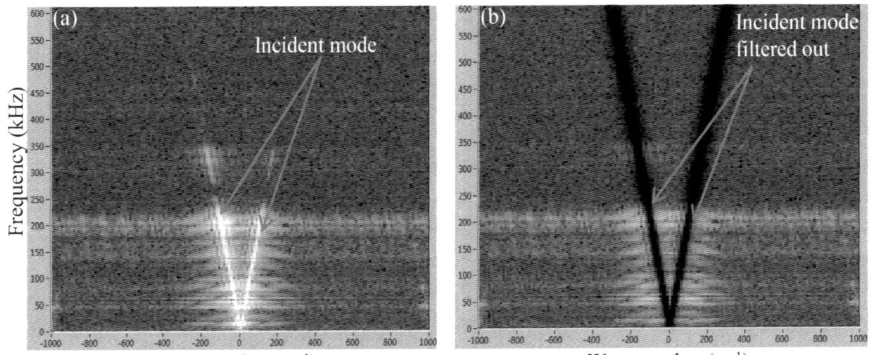

Figure 1. Frequency-wavenumber map (a) before and (b) after filtering of incident wave mode

Figure 2 shows intermediate inspection results after each step of data processing. Photograph of the specimen is given in Figure 2(a). It shows charring within the thermal damage infliction area that is indicated with a dotted circle. A snapshot from the ultrasound wavefield propagation imaging (UWPI) [2] video before and after the filtering of incident mode is given in Figure 2(b) and (c), respectively. The anomalous waves caused by the thermal damage is obscured by high amplitude of incident waves, hence unnoticeable in Figure 2(b). Once the incident mode was filtered out, the anomalous waves can clearly be seen in the middle of Figure 2(c), despite polluted by background noises. Further processing the mode-filtered data into an amplitude map enhanced the visibility of the damage, as shown in Figure 2(d). Note that the amplitude map facilitates easy detection of the damage, but the size of the damage could not be measured directly since the boundary of the damage could not be clearly defined yet, due to the background noises.

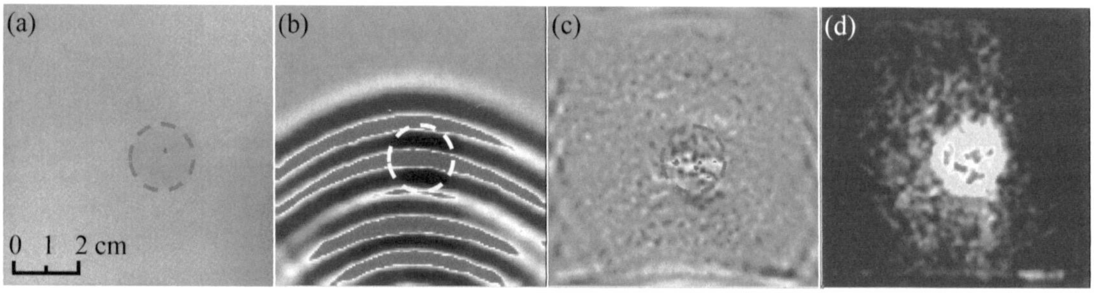

Figure 2. Intermediate inspection results based on (a) naked eye inspection; UWPI video snapshots (b) before and (c) after filtering of incident wave mode; and (d) amplitude map of anomalous waves without damage thresholding.

4.2. Effects of damage thresholding

Statistically Thresholded Anomaly Mapping (STAM) result images will be produced after applying the damage threshold. There are two variables affecting the results. The first variable is the value of N, which physically represents the number of samples used for the calculation of damage threshold. The value of this variable is more conveniently determined as P, which is a percentage of the total pixel number in an amplitude map. Applying the damage threshold calculated based on a constant value of $M = 4$ but varying the value of P from 10% to 50% gives interesting results in Figure 3. When a higher P is used, the background noise in the STAM image is significantly reduced. This is because, as more noise samples are gathered, the mean and standard deviation of background noise are closer to their true values. Reduction of background noises is important to allow the users to be able to distinguish the damaged area from the pristine area without a doubt regarding the actual damage size depicted by the STAM image. Concluding from these results, $P = 50\%$ is preferred as it has the least background noises. Setting P to a value higher than 50% is not preferred as users wouldn't know the damage size beforehand in real life cases, such that it poses a possibility of regarding a damaged area as noise and subsequently remove it from the final result, which in turn causes underestimation of the real damage size.

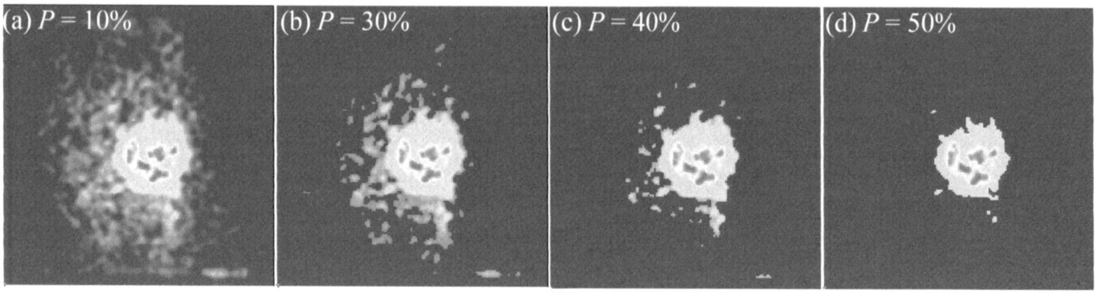

Figure 3. STAM images after applying a damage threshold calculated based on a constant $M = 4$ and different statistical sample size P at (a) 10%, (b) 30%, (c) 40%, and (d) 50%

The second variable affecting the results is the multiplier M in Eqn. 1. STAM images after applying the damage threshold calculated based on a constant value of $P = 50\%$ but varying the value of M from 1 to 8 are given in Figure 4. According to the Empirical Rule, 68% of data, which in this case is the background noise, falls within 1 standard deviation of mean, 95% of data falls within 2 standard deviations of mean and finally, 99.7% of data falls within 3 standard deviations of mean. By using $M = 4$, it is assumed that 99.9% of background noise will be covered, which means that the result will be noise free with a 99.9% confidence. Since $M = 4$ will result in a 99.9% removal of background noise, it can be concluded that it is the best choice as it is redundant to go higher than that. Other than that, a higher multiplier poses a threat of setting the damage threshold value too high such that even the damaged areas would be classified as background noise. This is shown in the Figure 4(d) where the damage could be measured to have a 20 mm diameter, even when the thermal damage can be safely assumed to be larger than that, simply because heat may be conducted to the surrounding of actual contact area with the 20 mm diameter heat damage inflicting aluminium rod. This proves that when $M = 8$ is used, only the most damaged area will be considered as damage while the less damaged area is considered as noise.

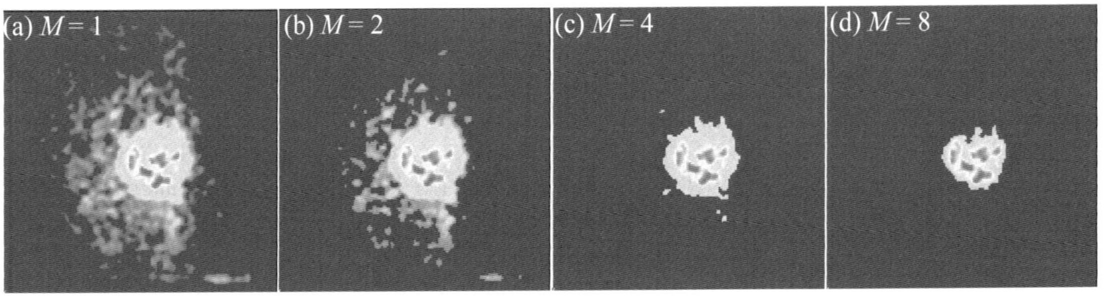

Figure 4. STAM images after applying a damage threshold calculated based on a constant $P = 50\%$ and different values of multiplier when (a) $M = 1$, (b) $M = 2$, (c) $M = 4$, and (d) $M = 8$.

5. Concluding remarks

The inability to statistically evaluate the damage size based on amplitude mapping has long plagued the UPI system. This paper proposed a damage visualization method called the STAM method to solve this problem. The STAM can isolate anomalous waves that are mode-converted from incident waves-damage interaction, and then map their amplitudes into an image in which the existence and location of damages can be visualized. The boundaries of damages can be identified by utilizing the statistical damage threshold calculation component of STAM, which facilitates accurate damage size evaluation.

Experimental validation of the proposed STAM method was conducted by inspecting a glass fibre reinforced polymer specimen inflicted with a thermal damage. A UPI system with scanning laser as ultrasound exciter and a fixed broadband ultrasound sensor was used to acquire the ultrasonic full wavefield data. Intermediate result in the form of anomalous waves amplitude map showed that the thermal damage could be clearly visualized, although the map suffered from some background noises. The effect of the statistical damage threshold calculation component of the STAM was examined by varying two parameters independently, i.e. M, the standard deviation multiplier, and P, the number of samples used for the calculation of damage threshold in term of percentage of total pixel number in the amplitude map. The results showed that $P = 50\%$ and $M = 4$ is the best combination to accurately evaluate the shape and size of the damage while eliminating 99.9% of background noises.

The proposed STAM method proved to be able to accurately locate and determine the size of a damage as long as anomalous waves are present in the ultrasonic wavefield data. It has the advantage of not requiring a reference data from a pristine specimen or a pristine region of the same specimen. Furthermore, its functionality and accuracy are not dependent on the correct selection of any specific wave mode, hence circumvented the need of performing reference inspection or numerical study of pristine specimen for the understanding of the complex ultrasonic waves and modes. The statistical damage threshold calculation component of STAM is suitable for other anomalous waves amplitude mapping applications as well, provided that damaged area is smaller than $P = 50\%$ of imaging area.

The authors are currently improving the STAM method by optimizing the incident mode filter so that even the incident waves at the edges can be filtered out fully. Quantitative accuracy and reliability analysis are also underway.

Acknowledgement
This work was supported by the Universiti Putra Malaysia through research project GP/2017/9564300.

References

[1] Kouche A E, Hassanein H S. Ultrasonic non-destructive testing (NDT) using wireless sensor networks. Procedia Comput Sci 2012; 10:136–43
[2] Chia C C, Lee J R, Park J S, Yun C Y, Kim J H. New design and algorithm of ultrasound propagation imaging system using Q-switched pulsed laser. 38th Int Conf Defektosk, 2008
[3] Chia C C, Jeong H-M, Lee J-R, Park G. Composite aircraft debonding visualization by laser ultrasonic scanning excitation and integrated piezoelectric sensing. Struct Control Heal Monit 2012; 19: 605–20
[4] Chia C C, Lee J R, Park C Y. Radome health management based on synthesized impact detection, laser ultrasonic spectral imaging, and wavelet-transformed ultrasonic propagation imaging methods. Compos Part B Eng 2012; 43: 2898–906
[5] Lee J-R, Chia C C, Park C-Y, Jeong H. Laser ultrasonic anomalous wave propagation imaging method with adjacent wave subtraction: Algorithm. Opt Laser Technol 2012; 44:1507–15
[6] Chia C C, Lee J-R, Park C-Y, Jeong H-M. Laser ultrasonic anomalous wave propagation imaging method with adjacent wave subtraction: Application to actual damages in composite wing. Opt Laser Technol 2012; 44:428–40
[7] Michaels T E, Ruzzene M, Michaels J. Incident wave removal through frequency-wavenumber filtering of full wavefield data. AIP Conf Proc 2009; 1096: 604–11
[8] Michaels T E, Ruzzene M, Michaels J E. Frequency-wavenumber domain methods for analysis of incident and scattered guided wave fields. Heal Monit Struct Biol Syst 2009; 7295: 729513-729513–12
[9] Alguri K S, Chia C C, Harley J B. Model-driven, wavefield baseline subtraction for damage visualization using dictionary learning. In: Chang F, Kopsaftopoulos F, editors. Proc. 11th IWSHM, vol. 2, Stanford, California, USA: DEStech Publications, Inc; 2017, p. 2276–83
[10] Larsson A. The interaction between a lightning flash and an aircraft in flight. Comptes Rendus Phys 2002; 3:1423–44
[11] Gagné M, Therriault D. Lightning strike protection of composites. Prog Aerosp Sci 2014; 64:1–16
[12] Hirano Y, Katsumata S, Iwahori Y, Todoroki A. Artificial lightning testing on graphite/epoxy composite laminate. Compos Part A Appl Sci Manuf 2010;41: 1461–70
[13] Capriotti M, Kim H E, Di Scalea F L, Kim H. Detection of major impact damage to composite aerospace structures by ultrasonic guided waves and statistical signal processing. Procedia Eng 2017; 199:1550–5
[14] Lamb H. On waves in an elastic plate. Proc R Soc London Ser A 1917; 93:114–28
[15] Rose J L. Ultrasonic waves in solid media. Cambridge, UK: Cambridge University Press; 1999
[16] Pieczonka Ł, Ambroziński Ł, Staszewski W J, Barnoncel D, Pérès P. Damage detection in composite panels based on mode-converted Lamb waves sensed using 3D laser scanning vibrometer. Opt Lasers Eng 2017; 99:80–7
[17] Ruzzene M. Frequency–wavenumber domain filtering for improved damage visualization. Smart Mater Struct 2007; 16:2116–29
[18] Michaels T E, Michaels J E, Ruzzene M. Frequency-wavenumber domain analysis of guided wavefields. Ultrasonics 2011; 51:452–66
[19] Flynn E B, Chong S Y, Jarmer G J, Lee J-R. Structural imaging through local wavenumber estimation of guided waves. NDT E Int 2013; 59:1–10

IOP Conf. Series: Materials Science and Engineering **405** (2018) 012007 doi:10.1088/1757-899X/405/1/012007

Implementation of Swiss Cheese for UniKL MIAT hangar

Muhamad Syazwan Mat Ghani* and Wong Zheng Yi**

University Kuala Lumpur- Malaysian Institute of Aviation Technology, 43900 Selangor, Malaysia

*msyazwanmg@unikl.edu.my, **wong.zheng@s.unikl.edu.my

Abstract. The aim of this review paper is to ensure the safety of aircraft maintenance personnel and students in hangar during ground handling by proposing the appropriate safety measures. The aviation organizations had put many efforts for the development and improvement of safety performance within these years to have a safety culture. The aircraft maintenance technicians need to have detailed knowledge of safety and risks that happen in hangar. This paper presents Swiss Cheese method that is proposed to be implemented at UniKL-MIAT hangar to improve the safety conditions.

1. Introduction

The impression that 'accidents are unavoidable' is strongly related to the aviation works that are high-tech and risky [1]. The connection between organisation's safety management processes and accident had been motivated due to the accident investigation reports from 1980 to 1990 [2-4]. The ways that can affect the conduct and dependability of safety systems are organisational practices where safety is managed in the aviation organizations, leading to either 'good' or 'lax' safety culture [5]. The care and concern of shared attitudes throughout the organization, and the commitment of observations by senior organization to safety contribute to the safety culture [6-7]. Safety has been defined as 'a state in which hazards and conditions leading to physical, psychological or material harm are controlled to preserve the health and well-being of individuals and the community' [8-9]. This is to avoid from any intentional or unintentional injuries, which should converge towards perception of protection from any danger.

The objective of this study is to ensure safety of the maintenance personnel and also students during the ground handling in the hangar, particularly the identification on how to reduce human error in the hangar during aircraft servicing and maintenance works. This is also conducted to increase the students' knowledge on the importance of safety by following the rules and also regulation while working in the aircraft hangar. The study is conducted through survey questionnaire that has been distributed to three categories of respondents: students, technicians and lecturers in University Kuala Lumpur-Malaysian Institute of Aviation Technology (UniKL-MIAT). The analysis of the responses from the questionnaire is taken into account to the recommendation made in this project.

2. Literature review

Although the aviation field is highly regulated and is the safest means of transportation, there are still accidents constantly happened and this can be disturbing and worrisome to several groups at any time. The supervision and cooperation of aviation regulators, air operators and service providers is necessary for the continued processing of a safe air transportation system within the complex and dynamic aviation environment [10]. Literature review done in this study provides some basic guidance in understanding the human error in hangar, including the impression of high hazard and concepts for high dependability, and the demonstration of the vitality of security in hangar. Issues of the human factors which influence the individual, working environment and tasks performed are also incorporated in this section [11].

The collaboration between people and machines understanding started as a system of ergonomic employment outline in the heavy industry. The three important segments for a security culture that have been examined are awareness, commitment and competence. This sense of duty regarding security must originate from top administration and work through the whole association [12, 13]. This notion has been repeated through with many risk management programs, which likewise require significant assistance from the masses. In security societies, it is required that the skills of people to be legitimately prepared and their tasks are proficiently performed. Lastly, the community within the system or process should be aware of the safety recommendations for any action and its subsequent effects to ensure the practice of the safety culture [14]. It has been asserted that 'awareness is the most crucial contributor to safe practice' [15]. This awareness is important in aircraft working environment because risk management depends on risk awareness to recognize things that are required to be handled. Practicing risk awareness in aircraft maintenance can lead to safety practices. Furthermore, the compliance to the requirements of the industry and federal regulation will facilitate the improvement in the practices of maintenance safety programs [13]. Development of maintenance CRM-type courses might help the understanding on the importance for the maintenance tasks to be carried out with minimum errors for the operational team [16].

2.1. Annex 19

Annex 19 contains the standard and recommended practices (SARPs) related to primary responsibilities and processes in relation to the safety management by the States. It was first accepted by the Council on 25 February 2013 in pursuant to the requirements of Article 37 of the Convention on International Civil Aviation (Chicago 1944) and designated as Annex 19 to the Convention [17]. The objective of proposing the SARPs in Annex 19 is to support the States in controlling the risks of aviation safety. Annex 19 still helps sustained development of a proactive strategy in improvising safety performance due to the high involvement of the global air transportation system to ensure safe operation in aviation activities. The basis of this proactive safety strategy is founded on the implementation of a State Safety Program (SSP) that addresses risks of safety systematically 18, 19].

SSP implementation will take time to be entirely completed and this could be affected by local and global air transportation system involvement and the realization of aviation safety oversight capabilities of the States [20]. The sources not only come from the current annexes regarding SSP and Safety Management Systems (SMS), but the Annex 19 is also a combination of safety usage of data elements and State safety oversight events. The advantages of combining these sources into a single annex in concentrating on the importance of consolidating the safety management activities and speed up the development of safety management provisions [21].

2.1.1. Safety Management System (SMS)

Certain state safety management roles required in Annex 19 might be represented to a regional safety oversight organization or a regional accident and incident investigation organization on behalf of the State [22]. Safety Management Systems as defined by ICAO is a method designed to manage and also conduct safety, including the structuring of organization, accountabilities, procedures and policies. This identifies the measurement of accepted level besides minimizing the risk in task, explaining the process that is applied to maintain safety [23]. A safety management system is a professional way to deal with security where an efficient, unequivocal and thorough process for overseeing dangers is implemented. Likewise, with all management system, a safety management system accommodates objective setting, arranging and estimating execution. By implementing the safety management system into the working structure of the association or organization, it will lead to a habitual safety practices by individuals while carrying out their tasks [24]. The adequacy of a safety management system relies upon the level of the engagement in the structure and practices of the association or organization on how the job or task has been carried out with the aim that a good safety and security culture is developed and monitored continuously [25]. Furthermore, SMS is an incorporated security framework by the top administration to account for all task safety techniques and dealing with a safe environment in a specific standard for the entire association to embrace in everyday activity [26].

Aircraft maintenance technicians give parts of their career to ground servicing and operating aircraft in hangar. They are also required to master the operation of the ground support equipment. Moreover, the maintenance technicians must have an exhaustive information of safety procedures used in aircraft servicing for the involvement of support equipment and the possible dangers present in aircraft hangar [27, 28]. Aircraft maintenance personnel should learn the importance of human factors to let them know how it influences the maintenance tasks done in hangar. Several areas could be considered such as Dirty Dozen even though many issues included when dealing with human performance [29]. Maintenance technicians must be made aware of how human factors can influence their working performance and the importance of safety in carrying tasks while doing maintenance practices in hangar [30].

3. Swiss Cheese method for UniKL MIAT hangar

Defences, barriers and safeguards occupy a key position in the system approach. The high technology systems have many defensive layers: some are engineered to rely on people while some others depend on procedures and administrative controls. The main target is to protect potential victims and assets from local hazards, which normally carried out very effectively but there are always will be weaknesses [31]. In an ideal world, each defensive layer would be intact. However, they are more like the slices of the Swiss cheese, having many holes—though unlike the cheese, these holes are continually opening, shutting and shifting their location. The presence of holes in any one "slice" does not normally cause a bad outcome. Usually, this happens only when the holes in many layers momentarily line up to permit a trajectory of accident opportunity—bringing hazards into damaging contact with victims, as shown in Figure 1.

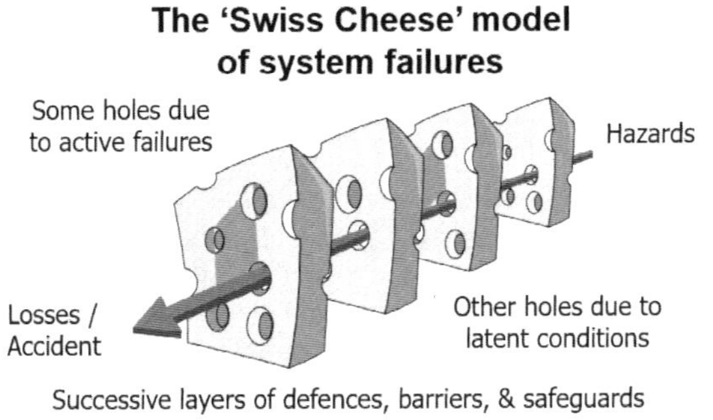

Figure 1. The Swiss Cheese model

Active failures and latent conditions are the reasons for the creation of the holes in the defences. A combination of these two sets of factors is normally involved in all adverse events. Active failures are the unsafe acts committed by people who are in direct contact with the patient or system, which could be in various forms such as slips, lapses, fumbles, mistakes and procedural violations. This kind of failures have a direct and typically short-lived impact on the integrity of the defences. For example, the technicians wrongly violated the maintenance procedures and skipped off few safety procedures, which immediately sparks high voltage of energy towards him that could lead to death. Followers of the person approach usually look no further for the causes of such adverse event once they have identified these proximal unsafe acts. Anyhow, as explained above, it could be seen that the causal history from such kind of acts can be clearly observed from slices of time to time of events. Meanwhile, latent conditions are the inevitable "resident pathogens" within the system. These could come from decisions of design engineers, technical service engineers, procedure writers and top-level management. Such decisions may be mistaken but they need not be. This mistaken decisions will eventually and potentially injecting pathogens into the system. Two kinds of adverse effect from latent conditions that will arise are: they can translate into error provoking conditions within the local workplace and they can create long-lasting holes or weaknesses in the defences. The pathogens may lie dormant within the system for many years before they combine with active failures and local triggers to create an accident opportunity. Latent conditions could be looked up, identified and solved before an adverse event occurs, unlike the active failures, which specific forms are normally hard to foresee. Understanding this leads to proactive rather than reactive risk management.

After some extremely discussion, the previous study from various researches have been reviewed and finalized according to the objectives. It is found that aviation industry is highly regulated industry with the safest means of transportation, but it is also at constant risk of major accidents that could happen anytime. Therefore, to achieve the safety of aircraft maintenance and students in hangar during ground handling and aviation personnel and to identify how to reduce human error in hangar during servicing of an aircraft, it is recommended from this study that the Swiss Cheese is the best choice for the implementation for UniKL MIAT hangar. A further detailed study on its implementation and also effectiveness is suggested for future work after this initial explorative study.

References

[1] Perrow C. Normal Accidents. New York: Basic Book, 1984
[2] Vette G. Impact Erebus. New Zealand: Hodder and Stoughton, 1983
[3] Wellington. Report of Royal commission to inquire into the crash of Mount Erebus, Antarctica, of a DC10 Aircraft. Air New Zealand Limited, New Zealand, 1981
[4] Moshansky V. Commission of inquiry into the air Ontario crash at Dryden,Ontario. Minister of Supply and Services Canada 1, Canada, 1992
[5] Reason J. Achieving a safe culture,theory and practice. Work and Practice 1998; 12: 293-306
[6] Pidgeon N O M. Organisational safety culture: Implications for aviation practice. In: McDonald N, Johnston N, Fuller R. (Eds.), Application of Psychology to the Aviation System, England: Aubury, 1995
[7] Soekkha H M. Aviation safety: Human factors, system engineering, flight operations, economics, strategies, management. Amsterdam: VSP, 1997
[8] Landry M, Bouchard M, Levaque C. Pratiques de sécurité urbaine dans le quartier Saint Roch de la ville de Québec, Québec," Préparé pour le Conseil de quartier Saint-Roch.Québec, 2000
[9] Kellermann A, Rivara F, Lee R, Banton J. Injuries and deaths due to firearms in the home. J Trauma, Injury, Infect Critical Care 1998, 45(2): 263-267
[10] Dannatt R. Role and effectiveness of government regulation of air transport safety. Australia: University of Western Australia, 2002
[11] Manoj S P, Jeffrey P, Melinda D T. Safety ethics: Cases from aviation, healthcare and occupational and environmental health, Burlington: Ashgate, 2005
[12] Rodenberg H, Blumen I. Air medical physician handbook. Salt Lake City: Air Medical Physician Association, 1999
[13] Kinnison H A. Aviation maintenance management. New York: McGraw-Hill, 2004
[14] Enoma N A. Developing key performance indicators for airport safety and security: A study of three Scottish airports. Edinburg: Herriot-Watt University, 2008
[15] Freiwald D, Lenz-Anderson C, Baker E. Assessing safety culture within a flight training organization. Journal of Aviation/Aerospace Education and Research 2013; 22(2)
[16] Hobbs A, Williamson A. Aircraft maintenance safety survey results. Australian Transport Safety Bureau, Australia, 1998
[17] Chien-Tsung L, Bos P, Caldwell W. System safety application: Construcitng a comprehensive aviation system safety management model. International Journal of Applied Aviation Studies 2007; 7(1): 28-35
[18] Wood R H. Aviation safety programs. Washington: A Management Handbook, 1991-1997
[19] Department of Transportation. Safety management system for airports [Online]. http://www.faa. gov/documentLibrary/media/
[20] Russell P. Management strategies for accident prevention. Air Asia, 1994
[21] Byron B. Practical safety management. Flight Safety Australia, 2001
[22] I. Annex 19. Safety Management. International CIvil Aviation Organisation, 2010
[23] Stolzer A, Halford C. Implementing safety management systems in aviation. Netherlands: Ashgate Publishing, 2012
[24] Canada T. Aviation safety management: Introducing a safety approach to safety management. Canada: Transport Canada, 2001
[25] Gill G, Shergill G. Perceptions of safety management and safety culture in the aviation industry in New Zealand. Journal of Air Transport Management 2004; 10: 233-239
[26] CAA. Safety management systems for commercial air transport operations. UK: Civil Aviation Authority United Kingdom, 2002
[27] AFS-600. Inspection Fundamentals. Aviation Maintenance Technician Handbook, 2008

[28] Department for Business. UK aerospace maintenance, repair, overhaul and logistics industry analysis. UK Government Department for Business, Innovation & Skills, 2017
[29] Nagel D. Human error in aviation operations. San Diego: Academic Press, 1988
[30] Avers K, Johnson B, Banks J, Wenzel B. Technical documentation challenges in aviation maintenance. A Proceedings Report, Washington DC, 2012
[31] Reason J. Managing the risks of organizational accidents. Aldershot: Ashgate, 1997
[32] Johnston N R N. Aviation psychology in practice. Aldershot: Ashgate, 1994
[33] Hawkins F H. Human factors in flight. Aldershot: Ashgate, 1993
[34] Wiegmann D A, Shappell S A. A human error approach to aviation accident analysis: The human factors analysis and classification system. United States : Routledge, 2017

IOP Conf. Series: Materials Science and Engineering **405** (2018) 012008 doi:10.1088/1757-899X/405/1/012008

Measuring variable amplitude loading with fibre optic

Wahyu Kuntjoro[1,2*], Ramzyzan Ramly[1,2] and Najmuddin Assanah[2]

[1] Aviation Technology Research, Universiti Teknologi MARA, 40450 Shah Alam, Selangor, Malaysia

[2] Flight Technology and Test Centre (FTTC), Faculty of Mechanical Engineering, Universiti Teknologi MARA, 40450 Shah Alam, Selangor, Malaysia

*wkuntjoro@yahoo.com

Abstract. Cyclic loading is known to be responsible for fatigue failure. The variable amplitude loading is a general type of fatigue load and the most common one in practice. Monitoring of fatigue loading such as variable amplitude loads is commonly conducted using strain gauges. The use of fibre optics like fibre Bragg gratings (FBG) is attractive because this type of sensor is not easily interfered by the electromagnetic fields. Its suitability for measuring accurately the cyclic loadings need to be investigated. This paper is reporting the experiments to investigate the suitability of FBG to measure variable amplitude loadings. FBG test setup using a standard fiber optic measurement system was prepared and the cyclic test was conducted using universal testing machine. The fatigue test specimens used were prepared according to the ASTM E606 standards. The obtained result shows that FBG is able to measure variable amplitude loads and this means that FBG can potentially be used for fatigue monitoring of structures.

1. Introduction

Fatigue of structures happens due to cyclic loads and this can lead to catastrophic failure. This type of failure is dangerous as it happens suddenly. A way to avoid fatigue failure is by continuous monitoring of the structure, hence the term structural health monitoring (SHM). The structural deterioration over time due to fatigue can be predicted to calculate the remaining life by analyzing the gathered history data through SHM program [1]. Development of a fatigue monitoring to fighter aircrafts was reported by Kuntjoro et al. (2009) [2]. The analysis was based on Low Cycle Fatigue (LCF) and the monitoring was based on the accelerometer and strain-gauges that were installed on the airplanes.

To perform fatigue analysis, fatigue parameters must be known. One important parameter is fatigue load, in the form of cyclic load spectrum [3]. To obtain the fatigue load spectrum, a good sensor shall be attached to the structure. A sensor must be sensitive to the measured parameters, does not influence the behavior of structure, durable and sensitive to small changes. Strain gauges are normally applied for this purpose. However, the use of fiber optics such as fiber Bragg gratings (FBG) is attractive as this type of sensor is not easily interfered by the electromagnetic fields. FBG sensor is sensitive to the change of physical parameters by responding in the form of a reflected light wavelength shift in a very narrow bandwidth (typically 0.1-0.5 nm) [4].

Tahir et al. (2008) performed a comparison study between FBG sensor and electrical strain gauge for strain measurement [5]. They stated that their obtained FBG sensor data correlates better with the theoretical calculation. They found that FBG is immune to electromagnetic interface, can measure very high strain (>10,000 μm/m) and have good corrosion resistance. In another study, Li and Jiang (2009) also reported that FBG can be used to measure very high strain [6]. Guan et al. (2000) found that fiber optic sensor can be used to measure both strain and temperature simultaneously [7]. This was achieved by measuring the transmitted intensity and wavelength at one of the loss peaks. Li et al. (2004) reported that fiber optics application for SHM has begun to widespread in civil engineering applications [8]. The use of fiber optics for SHM was reported in buildings, piles, bridges, pipelines, tunnels and also dams. Moreover, a research on the application of optical FBG sensor for composite honeycomb sandwiched panel monitoring was reported by Ramly et al. (2012) [9]. The FBG was embedded inside the layup and the research was aimed to identify delamination on the face of the panel.

A research of studying the response of FBG under variable amplitude load is reported in this paper. The idea is to make use of the data further for fatigue analysis. In this research, a standard specimen (ASTM E606) was used. FBG sensors were attached to the specimen and test configuration was set up. Comparison with load cell reading (which comes from strain gauge sensor) was performed.

2. Fatigue loads

The fatigue load or fatigue stress is often in the form of a sinusoidal pattern. Figure 1 illustrates some of the various stress-time relations. Figure 1(a) illustrates a complete reversed cycle of stress where the mean stress for this case is zero. On the other hand, Figure 1(b) illustrates a cyclic stress in which the maximum stress, σ_{max} and the minimum stress, σ_{min} are not equal to each other, and both are in tension. Last but not least, Figure 1(c) illustrates a rather general stress cycle on a component that is subjected to periodic unpredictable loads.

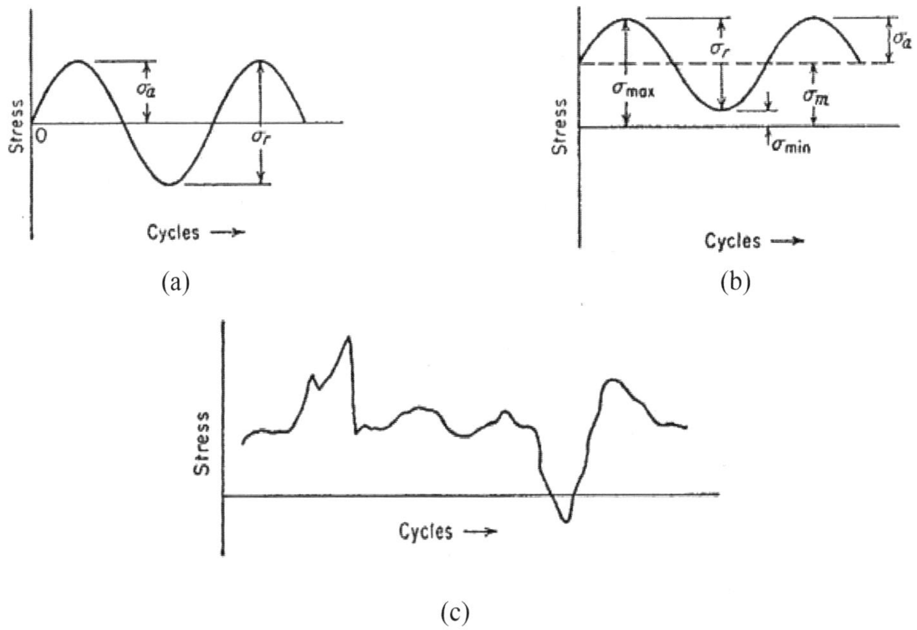

Figure 1. Cyclic stress

3. Fiber Bragg Grating (FBG)

Schematic diagram of FBG is shown in Figure 2. When light travels between the medium of different refractive indices, it may reflect and refract at the interface. Periodic perturbation of the core refractive index along the fibre length gives FBG the ability to work as a wavelength selective mirror. Bragg's wavelength, λ_B is defined at a condition when the reflectivity of the grating is maximum and is given as Eqn. 1, where n_{eff} is the effective index of the grating in the fibre core and Λ is the grating period.

$$\lambda_B = 2n_{eff}\Lambda \quad (1)$$

Under extension or compression, the grating spacing of FBG is changed and therefore causes a shift of the Bragg wavelength. The wavelength shift due to the strain is defined as given in Eqn. 2, where ε is the strain.

$$\Delta\lambda = (1 - p_e)\lambda_B \varepsilon \qquad (2)$$

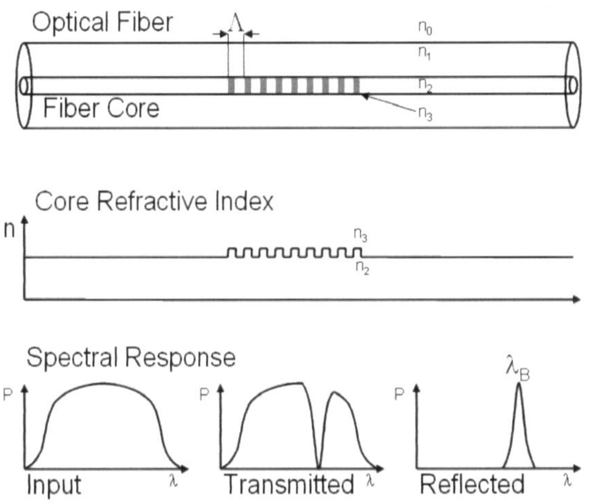

Figure 2. FBG schematic diagram [4]

The photoelastic coefficient, p_e is defined by Eqn. 3, where the $p_{I,j}$ coefficients are called the Pockel's coefficient of the strain-optic tensor and v is the Poisson's ratio [4].

$$p_e = \left(\frac{n_{eff}^2}{2}[p_{12} - v(p_{11} + p_{12})]\right) \qquad (3)$$

4. Cyclic load experiment

The test setup is shown in Figure 3. The equipment consists of FBG scanner, a notebook computer and the FBG itself as the sensor. The FBG was attached to the specimen. Variable loading was applied to the standard specimen (ASTM 606) to study the response from FBG sensor.

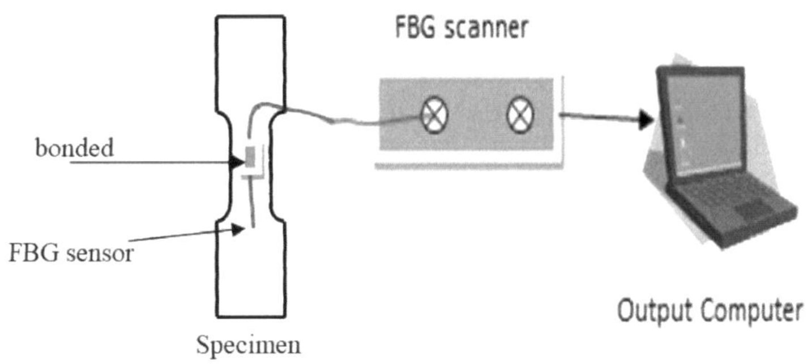

Figure 3. The setup of the FBG measurement system

The specimen was tested using Universal Testing Machine available in the Faculty of Mechanical Engineering, Universiti Teknologi MARA, Shah Alam, Malaysia. The setup was based on the variable amplitude loading, involving some sequences of constant amplitude cycles. The variable loading input command is reflected in the load cell readings. Three specimens with FBG sensors were manufactured but only one specimen produced the proper results. The FBG sensors of the other two specimens were damaged during testing. The results are shown in Figure 4 and Figure 5.

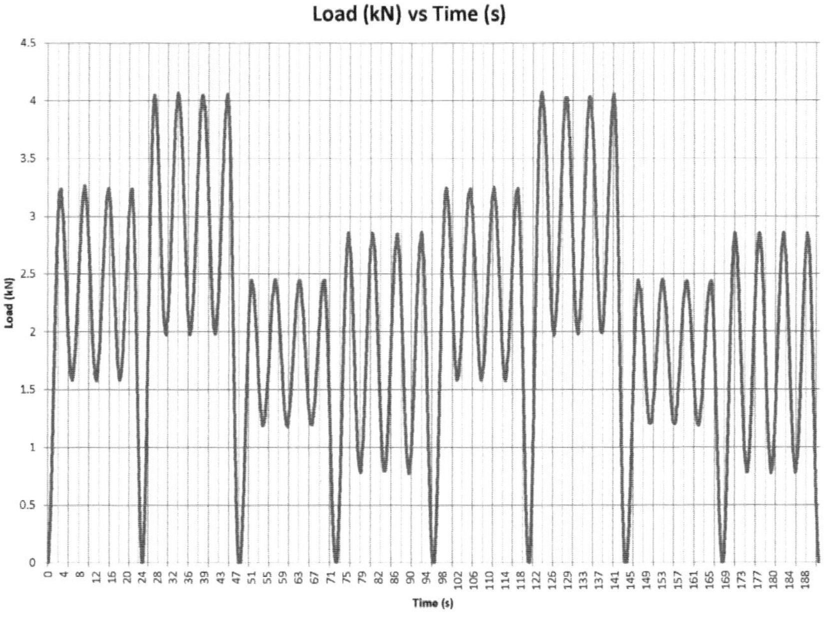

Figure 4. Variable loading applied to the standard specimen (load cell reading)

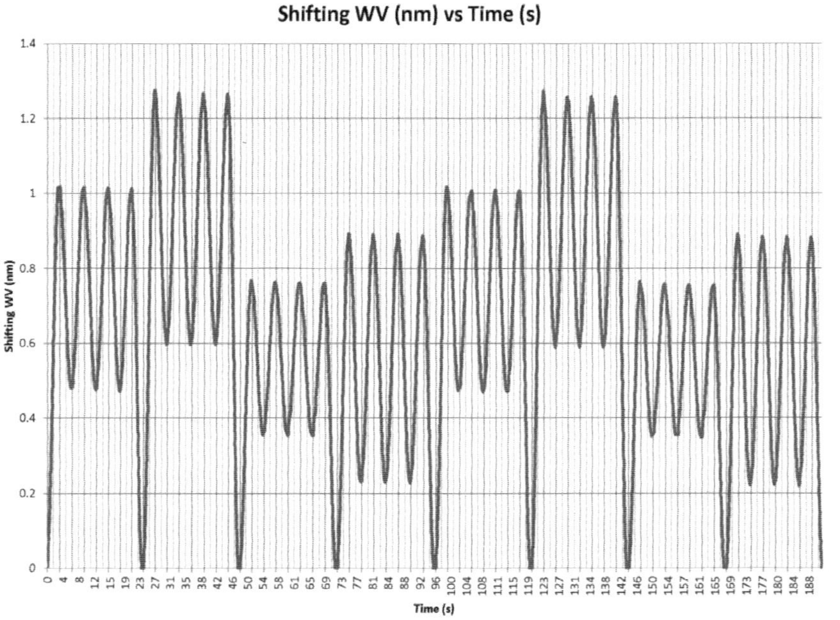

Figure 5. Response from FBG sensor

Figure 4 is constructed based on load cell reading whereas Figure 5 is plotted based on the FBG wavelength shift. From both Figures 4 and 5, it is clear that not only the shape of the graphs is similar but both graphs also have the same proportion. It means that response from FBG sensor is consistent with the variation loading applied. Strain analysis of the FBG output can be obtained from equations 1-3 and a study is being done for that purpose. Nevertheless, there is no conclusive result as yet. It is noteworthy that the strain is potentially better to be found based on a direct proportion/scaling between the load cell reading and the FBG wave length shift.

5. Conclusion

An investigation of FBG suitability for reading/sensing variable amplitude loading has been done. A fatigue specimen was prepared in accordance with ASTM 606. FBG fibre optics were attached to the specimens. A standard fibre optic test setup was prepared and the specimen was loaded under variable amplitude cycles using a Universal Testing Machine. The results show that response from FBG sensor is consistent with the variation loading applied. Hence it can be concluded that FBG is able to measure variable amplitude loads. There is a big FBG potential for application in aerospace engineering. It is especially attractive for application where the strain gauge is difficult or impossible to be used such as for sandwiched panel. However, this study also found out that FBG is fragile and needs to be carefully handled. The FBG sensor is better to be protected during tests by a kind of jacket, similar to jacket for cables.

References

[1] Balageas D. Structural health monitoring. ISTE Ltd 1st South Asian Edition, 2007
[2] Kuntjoro W, Ashari M S, Ahmad M, Mydin A M. Development of fatigue life monitoring of RMAF fighter airplanes. Book Chapter in ICAF 2009 – Bridging the Gap between Theory and Operational Practice, Springer.
[3] Budynas R G, Nisbett J K. Fatigue failure resulting from variable loading. Shigley's Mechanical Engineering Design Book, McGraw Hill, 2008
[4] Fernando G F. Fiber optic sensor systems for monitoring composite structure. RP Asia 2005 Conference, 2005
[5] Tahir B A, Saktioto J, Fadhali M, Rahman R A, Ahmed A. A study of FBG sensor and electrical strain gauge for strain measurements. Journal of Optoelectronics and Advanced Materials 2008; 10(10): 2564 – 2568
[6] Li S, Jiang D. Structural large strain monitoring based on FBG sensor. Symposium on Photonics and Optoelectronics, 2009
[7] Guan B O, Tam H Y, Tao X M, Dong X Y. Simultaneous strain and temperature measurement using a superstructure Fiber Bragg Grating. IEEE Photonics Technology Letters 2000; 12(6): 675- 677.
[8] Li H N, Li D S, Song D B. Recent applications of fiber optic sensors to health monitoring in civil engineering. Engineering Structures 2004; 26(11): 1647-1657
[9] Ramly R, Kuntjoro W, Abd Rahman M K. Using embedded Fiber Bragg Grating (FBG) sensors in smart aircraft structure materials. Procedia Engineering 2012; 41: 600-606

Generation of design alternatives of vertical passenger seat for standing cabin concept in commercial transport aircraft

Norhafizah Dasuki, Fairuz I Romli* and Mohammad Yazdi Harmin

Department of Aerospace Engineering, Faculty of Engineering,
Universiti Putra Malaysia, 43400 Serdang, Selangor, Malaysia.

fairuz_ir@upm.edu.my

Abstract. Standing cabin concept, whereby the passengers are transported during flight in their standing position, has been proposed as one of the measures to make flying more affordable to the mass public. The vertical seat, or also known as standing seat, is an essential element for the implementation of this new cabin concept as it will determine the compliance to the governing aviation regulations and also the level of safety and comfort for the passengers. Up until now however, there is no vertical seat design that has been approved for use inside the commercial transport aircraft cabin. Therefore, this study presents the generation of alternative designs for the vertical seat through the use of systematic engineering design methodology with the use of supporting design tools such as Quality Function Deployment (QFD) chart and Morphological Matrix. Five alternative concepts of the vertical passenger seat design have been developed for further assessment study to select the best one among them.

1. Introduction

The standing passenger cabin concept is one of the revolutionary solutions that have been proposed to increase the cabin capacity of commercial passenger transport aircraft in order to reduce flying cost per passenger. This idea has already been around for quite some time, with Spring Airlines, a low-cost airlines in China, been among the first to show high interest to implement the idea on its aircraft fleet back in 2006 [1]. In fact, a commercial pattern of what looks like a vertical stand for passenger cabin has been filed by Airbus, one of the main aircraft manufacturers in the world, in 2002 as depicted in Figure 1 [2]. In general, the standing cabin concept is readily in compliance with the current governing aviation regulations such as the ones imposed by the Federal Aviation Authority (FAA) and the Air Transport Association (ATA) that do not specify that the passengers must be seated during takeoff and landing phases, and do not impose any specific standard for seating comfort or arrangement inside the cabin [3]. A low-cost airline in Europe, Ryanair has been granted the regulatory approval to conduct a series of test flights with modified section of the passenger cabin on their aircraft as standing-only section in 2012 [4]. However, despite all these positive developments, there is still no standing cabin design that has been implemented or approved for commercial passenger transport aircraft until today.

There seems to be notable lack of published research study on this passenger cabin design concept and its implementation onboard of commercial passenger transport aircraft. In previous work done on this research topic, it has been highlighted that standing cabin concept can accommodate passengers with adequate comfort level [5] and its potential benefits to lower the flight cost per passenger is also discussed [6]. The results from the conducted public survey among Malaysian public have shown that cabin concept has notable market support and acceptance [7]. Furthermore, a conceptual design of the vertical seat for application in standing cabin concept has been proposed as in Figure 2 [8]. The work presented in this paper is a continuation of the previous research work, whereby the design of vertical seat to support and secure the passengers in standing cabin concept is further explored and improved from previous design.

Figure 1. Illustration of standing cabin in a pattern filed by Airbus [2]

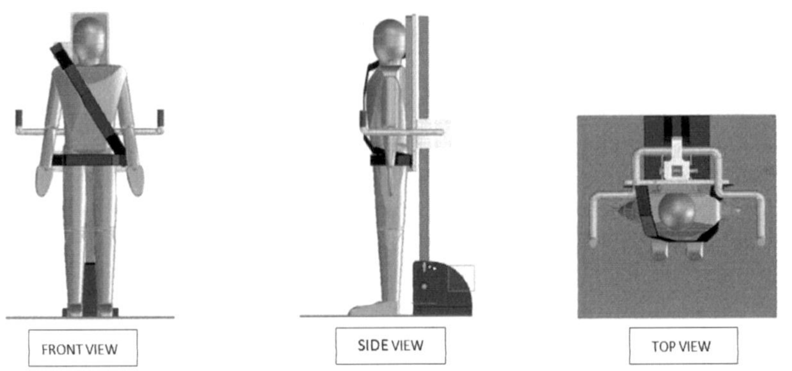

Figure 2. Proposed vertical seat design by Mohammad Nor and Romli (2013) [8]

2. Design alternatives generation process

Engineering product design process normally starts with the design requirement analysis [9]. In this step, the expectations and also preferences for the product to be developed with regards to its physical and functional aspects are identified and established. This is a very essential step that heavily dictates the probable success of the product design and development process because it determines whether the developed product design will match well with the market demands. There are several design aid tools that can be applied to aid the requirement analysis process and one of them is the Quality Function Deployment (QFD). QFD is a method that can be applied to satisfy the customers or stakeholders by translating their needs and preferences into design targets and quality assurance points for the product to be developed [10]. The main output of QFD method is a matrix known as House of Quality, which is essentially the conceptual map for the design process that establishes the underlying links between the customer requirements and the technical design parameters to satisfy them [11].

From the information summarized in the House of Quality, several essential design parameters and also functions of the product to be developed can be identified and focused on. Furthermore, from the information in the benchmarking section of the resultant House of Quality matrix, data on current competitor products in the market can be obtained and this can be utilized by product designers to set target performance and create market niche for the product under development. Using the information in QFD, another design aid tool called Morphological Matrix (MM) can then be applied to generate some alternatives for the product design. In short, MM is an assisting tool to help designers to explore the conceptual design space for the product, especially to encourage the search into new territories that are typically unexplored. Unlike other methods that can be heavily biased towards the designer's past experiences and also the previous or current successful designs in the market, MM can lead to more inventive process through random combinations of the possible solutions for each identified product attribute [12]. This allows the explored solution space to be extended outside the typical norm and may lead to the generation of new revolutionary ideas. However, each generated solution is assessed for feasibility and practicality before being considered further in the design and development process. MM has been accepted as a good means of recording the information on the possible solutions for the product design and facilitating the cognitive process of generating system-level design solution [13].

For this research study, summary of the design process applied up to the generation of alternative design concepts for the vertical seat is presented in Figure 3. A short public survey is conducted to the Malaysian public in order to obtain the customer requirements for the vertical seat design. In addition, interview sessions with few experts in related engineering industries, aviation authority and airlines are also done to gain some insights on their needs and requirements. All gathered information from the public survey and interview sessions has been summarized into the QFD matrix for the vertical seat as presented and discussed in Ref. [14], which is reproduced in Figure 4. Based on this information, MM for the vertical seat is constructed.

3. Generated design concepts

As previously mentioned, QFD is a well-accepted design tool that has been applied in many product design processes. From the data in the QFD shown in Figure 4, main design parameters and functions of the vertical seat for the standing cabin concept can be extracted. This information is then utilized to construct the MM for the vertical seat. All in all, six parameters or attributes for the vertical passenger seat design are identified: main seat structure, backrest, bottom seat pan, secure or safety mechanism, armrest and seat adjusting mechanism. Available options for each of them are identified, which have been obtained through literature review on existing designs and also from brainstorming process on new ideas and concepts. The constructed MM for the vertical passenger seat is shown in Figure 5.

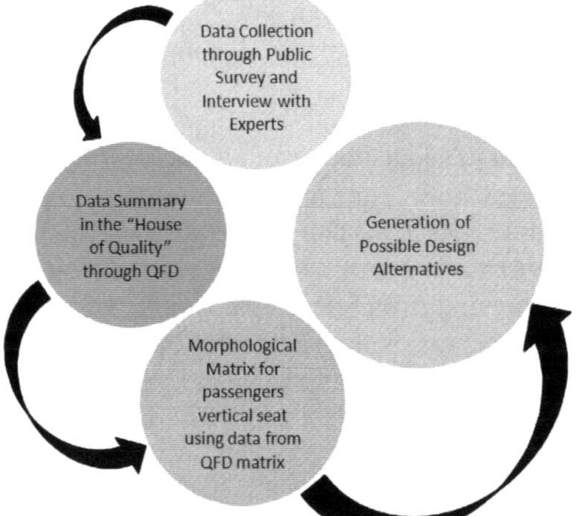

Figure 3. Design methodology flowchart for this study

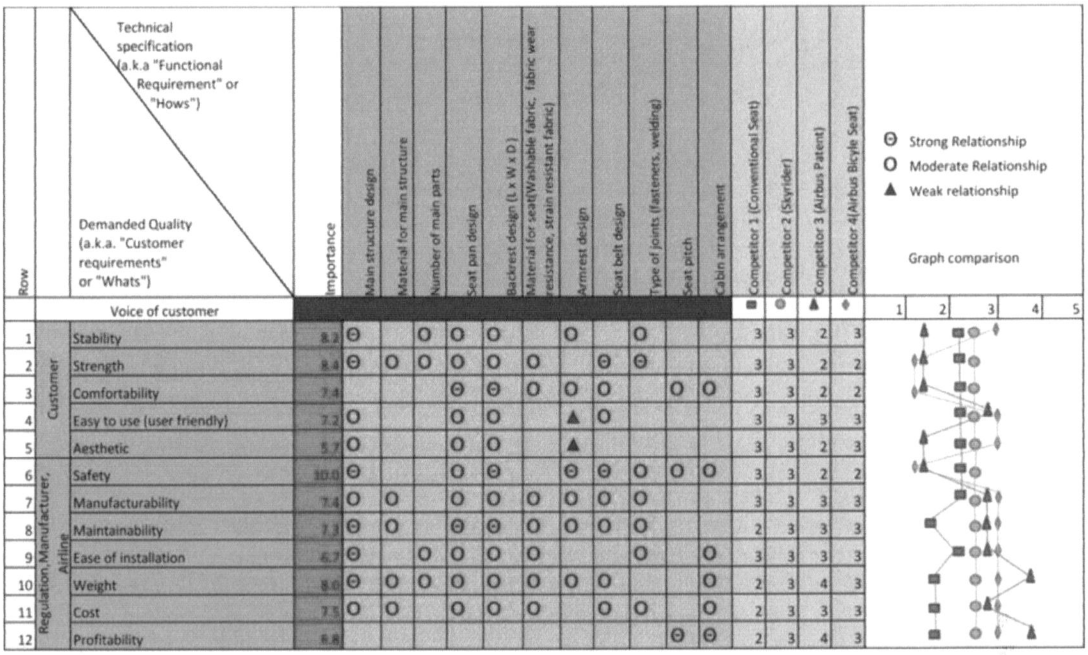

Figure 4. Constructed QFD matrix for the vertical seat [14]

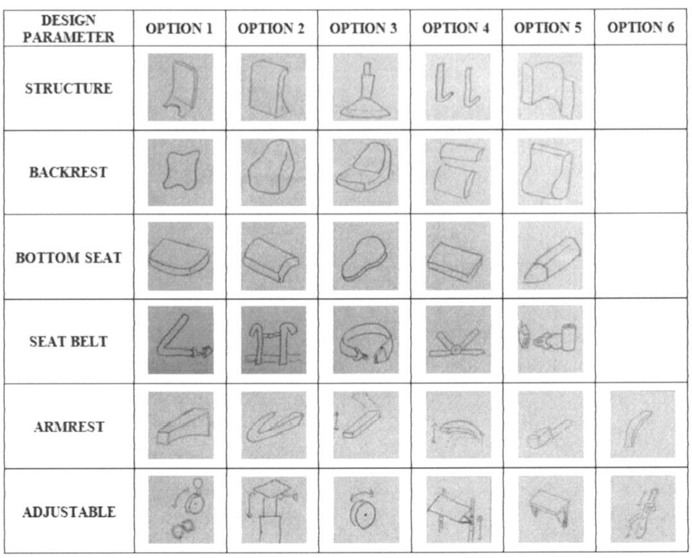

Figure 5. Constructed morphological matrix for the vertical passenger seat

Based on the MM in Figure 5, several design alternatives can be generated for the vertical seat. An explanation on how this is done can be found in Ref. [15]. The total number of all possible solutions that can be derived from the MM is the product of multiplying degree of freedom for each dimension and rows of listed attributes, which amounts to 22,500 possible different design combinations. There are many ways that have been proposed to deal with this so-called "curse of dimensionality" whereby the total number of combinations grows exponentially with each additional degree of freedom in the MM. In this study, the traditional random mix-and-match of the options for each attribute is applied to generate the design alternatives. Several initial design concepts are randomly generated before they are narrowed down to only five good designs to be seriously considered through a brainstorming process. This preliminary screening process is accomplished based on personal intuition and knowledge about existing designs.

3.1. Design Concept 1
One of the main driving factors for standing cabin concept is to increase the ability to accommodate more passengers in the same available cabin space. The vertical seat design in Concept 1, as shown in Figure 6, is selected for consideration to maximize number of passengers that can be accommodated inside the cabin space due to its design simplicity. Its main design structure uses a standard hollow cylindrical beam to reduce its overall structural weight. The seat's backrest is designed to support and provide comfort to the standing passengers. In addition, this vertical seat is also equipped with the adjustable seat pan and armrest to provide more comfort, particularly in relieving some of the body stresses caused by prolonged standing posture. Last but not least, a two-point safety belt mechanism is included for securing the passengers during flight.

3.2. Design Concept 2
Similar to the first concept, Concept 2 that is considered for design of the vertical passenger seat has also been driven by the notion of space-saving and more cabin accommodation capacity. The design concept is illustrated in Figure 7, which closely resembles a typical office chair. The main structure is a single cylindrical beam, with a hydraulic system installed within it for height adjustment capability. The seat pan and backrest is designed together as one unit, which means that the seat pan is fixed at a certain angle and is not adjustable. The adjustable armrests are provided to provide some comfort and support to passengers while they are standing. For safety mechanism, a three-point seat belt is installed to secure the passengers during flight.

3.3. Design Concept 3
The generated design Concept 3 for vertical passenger seat has been made primarily for high comfort and safety purposes. As can be observed in Figure 8, the primary seat structure looks bulky and rigid, providing strength and more stability to the seat. Apart from backrest and adjustable seat pan, it is also equipped with a headrest. The adjustable armrests are provided for additional support to the standing passengers and two-point seatbelt mechanism is installed to secure the passengers for safety purposes during flight. It should be noted that unlike the previous two concepts, this seat concept will be much heavier but stronger and steadier, which is expected to be able to provide more comfort (with less seat vibrations) and better safety support to the passengers. Moreover, with this seat concept, the number of standing passengers accommodated inside the cabin space will not be as many as those with previous two concepts.

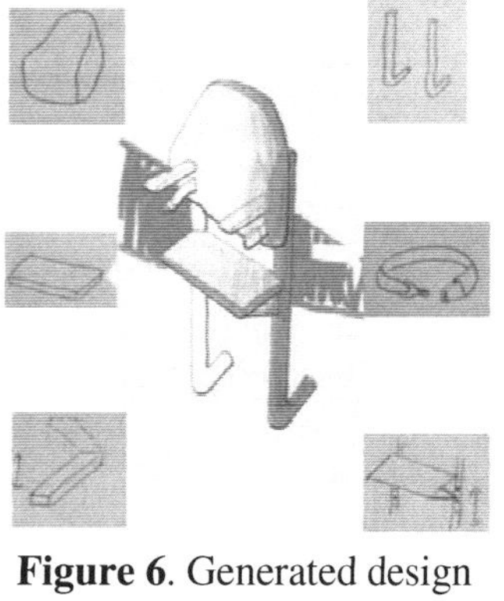

Figure 6. Generated design Concept 1 for the vertical seat

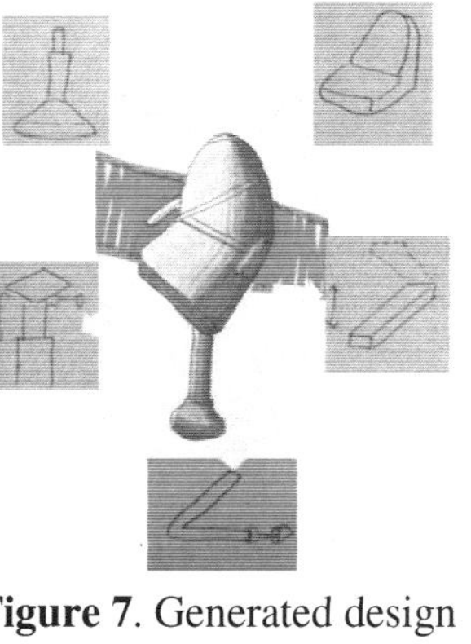

Figure 7. Generated design Concept 2 for the vertical seat

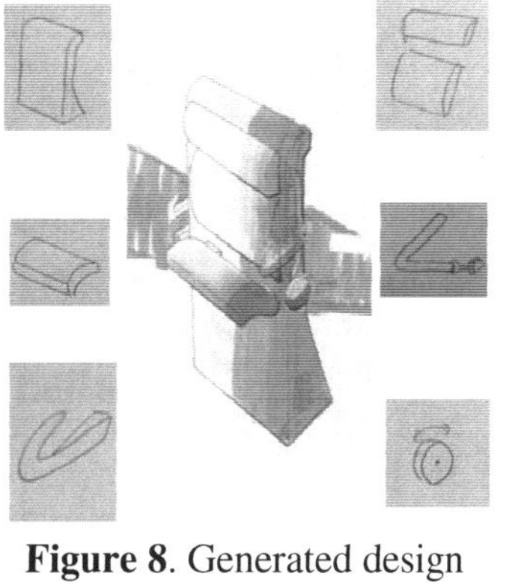

Figure 8. Generated design Concept 3 for the vertical seat

IOP Conf. Series: Materials Science and Engineering **405** (2018) 012009 doi:10.1088/1757-899X/405/1/012009

3.4. Design Concept 4

The main added element for the generated design Concept 4 for the vertical passenger seat is the sense of privacy. The main primary structure of the seat is shaped like a cocoon and instead of armrests, it is equipped with some sorts of a simple hand holder on both sides of the structure for the passengers to hold on for support during flight. Furthermore, a comfortable backrest and an adjustable seat pan are also equipped for the seat. A two-point seatbelt mechanism is also provided to secure the passengers. Similar to the previous Concept 3, this seat concept appears to be heavier and because of its structural shape, its manufacturing complexity and cost will be higher. Nevertheless, its advantages included the provision of more comfort to the passengers with the added privacy and less seat vibrations, with a stronger seat structure for safety. The use of cabin space to install this seat concept will also make it unable to fully maximize the available space for maximum number of standing passengers inside the cabin. The illustration of this concept is shown in Figure 9.

3.5. Design Concept 5

Last but not the least, the generated design Concept 5 of the vertical seat for the standing passengers looks more aesthetical than previous designs as depicted in Figure 10. It shares an almost similar main structural design like Concept 3 but it is lighter and curved. Similar to some of the previous concepts, it is equipped with a comfortable backrest, adjustable seat pan and also flexible armrests that can be adjusted up and down. Its safety mechanism is a two-point seatbelt to secure standing passengers. It is believed that this seat design can provide the level of safety and comfort much like previous Concept 3 and Concept 4, but with lighter weight and requires less space. The latter helps to add more possible accommodation of standing passengers in the cabin.

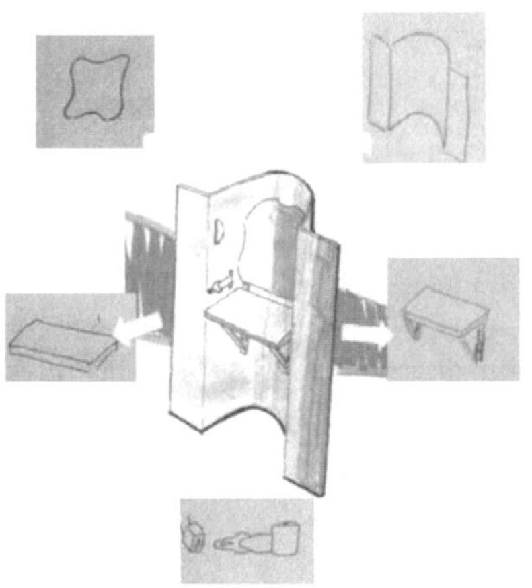

Figure 9. Generated design Concept 4 for the vertical seat

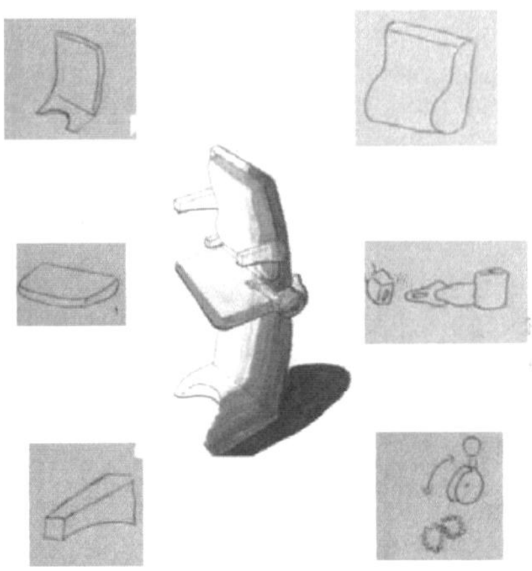

Figure 10. Generated design Concept 5 for the vertical seat

3.6. Further discussions

Once the design alternatives for the vertical seat have been generated, the next following step will be to evaluate them to select the best one among them. The evaluation criteria to be used in this process can be referred back to the established customer needs and requirements from previously constructed QFD chart for the vertical seat. By doing this, the chosen design concept for further development is ensured to be the closest one to meet most of the anticipated performance and characteristics by its stakeholders and end users. There can also be cases, especially with the purely randomize mix-and-match method from the MM, where none of the considered generated concepts is deemed suitable to be further developed based on evaluation results against its requirements. In such cases, the concept generation process from MM has to be reiterated and some new alternative design concepts are down selected and evaluated. This continues until one design concept emerges as the most potentially good one for further development.

As noted from the five generated design alternative concepts, each of them has their own strengths and weaknesses that are hard to differentiate without using a proper evaluation technique. In addition, established customer needs and requirements that become the evaluation criteria can be conflicting to each other. Hence, systematic multi-criteria decision making techniques and tools are often applied for the design evaluation process like Pugh Analysis or Technique for Order of Preference by Similarity to Ideal Solution (TOPSIS).

4. Conclusion

In this paper, few design alternative concepts of the vertical passenger seat for the standing cabin have been generated through a systematic design engineering process. Based on constructed QFD matrix for the vertical passenger seat in previous study, the information is utilized to construct the corresponding morphological chart. Five different design concepts are generated from the morphological chart by a random mix-and-match procedure. The next following step would be to evaluate these five concepts against the established customer requirements or needs in order to choose the best one among them for further development.

Acknowledgement

The authors acknowledge the financial support for this study from Universiti Putra Malaysia (UPM), Malaysia through research grant GP-IBT/2013/9407100.

References

[1] Pae P. Chinese airline may offer cheaper fares to passengers who stand during short flights. Los Angeles Times, 2009

[2] Angerani J, Baatz A, Hiesener S, Merensky H. US Patent No. 6,467,728 B2 (22 October 2002)

[3] Elliot C. One day, that economy ticket may buy you a place to stand. The New York Times, 2006

[4] Kerridge S. Ryanair–Approval for standing only cabin areas. http://blog.france-airport-guide.com/2012/04/ryanair-approval-for-standing-only.html [Accessed: December 2013]

[5] Romli F I, Asmadi A N, Dasuki N. Ergonomics study of vertical seat design for standing cabin concept in commercial transport aircraft. International Review of Aerospace Engineering 2015; 8(3): 17-22

[6] Romli F I, Mohamad Noor A R, Dasuki N. A study on potential standing cabin effects in improving the competitiveness of low-cost airlines. IACSIT Int. Journal of Engineering and Technology 2014; 6(4): 304–308

[7] Romli F I, Dasuki N, Harmin M Y. Market potential study for standing cabin concept for domestic low-cost commercial airlines in Malaysia. IOP Conference Series: Earth and Environmental Science 2016; 30: 012012

[8] Mohamad Noor A R, Romli F I. Conceptual design of vertical passenger seat for standing cabin in commercial transport aircraft. Journal of Advanced Management Science 2013; 1(4): 344-348

[9] Zeng Y, Gu P. A science-based approach to product design Part II: formulation of design requirements and products. Robotics and Computer Integrated Manufacturing 1999; 15: 341-352

[10] Akao Y. Quality function deployment: Integrating customer requirements into product design. Boca Raton: CRC Press, 2004

[11] Gonzalez M E, Quesada G, Bahill A T. Improving product design using quality function deployment: The school furniture case in developing countries. Quality Engineering 2003; 16(1): 45-56

[12] Arciszewski T. Morphological analysis in inventive engineering. Technological Forecasting and

IOP Conf. Series: Materials Science and Engineering **405** (2018) 012009 doi:10.1088/1757-899X/405/1/012009

Social Change 2018; 126: 92-101

[13] Weber R G, Condoor S S. Conceptual design using a synergistically compatible morphological matrix, 28th Annual Frontiers in Education Conference. Moving from 'Teacher-Centered' to 'Learner-Centered' Education, 1998

[14] Dasuki N M, Romli F I. Quality function deployment for new standing cabin concept of commercial transport aircraft. Journal of Mechanical Engineering 2018; SI5(2): 247-257

[15] Smith G, Richardson J, Summers J, Mocko G M. Concept exploration through morphological charts: An experimental study. Journal of Mechanical Design 2012; 134(5): 051004

IOP Conf. Series: Materials Science and Engineering **405** (2018) 012010 doi:10.1088/1757-899X/405/1/012010

Parametric optimization and performance analysis of a waste heat recovery system using Organic Rankine Cycle (ORC)

N H H Zainol Abidin, S Saadon* and A R Abu Talib

Department of Aerospace Engineering, Faculty of Engineering, Universiti Putra Malaysia, 43400 Serdang, Selangor, Malaysia.

* mimisaadon@upm.edu.my

Abstract. Organic Rankine Cycle (ORC) has been proven to be the most effective solution to utilize fully the waste heat energy produced from manufacturing and power plant. In this paper, thermodynamic analysis and performance optimization of an ORC waste heat recovery system is performed for an aircraft engine and R245-fa is chosen as the refrigerant. The main concern of this study is to maximize the total heat transfer rate through the evaporator, thus the optimization of the Thrust Specific Fuel Consumption (TSFC) and the Specific Thrust of the engine can be performed. The main parameter to conduct the optimization is Turbine Inlet Temperature (TIT) and preheating temperature of R245fa. From the analysis, by increasing the Turbine Inlet Temperature and the preheating temperature, the TSFC and Specific Thrust of the engine can be optimized.

1. Introduction

Organic Rankine Cycle (ORC) is widely known for its capability to convert low grade heat into some form of energy that can be used. The ORC can effectively recover the low grade heat source due to its distinctive thermodynamic performance. The history of internal combustion engine exhaust waste heat recovery has been reviewed in [1], focusing on ORC since this thermodynamic cycle works well with the medium-grade energy of the exhaust. The primary focus of the review is the selection of the cycle expander and working fluid since they have been regarded as having the largest impact on the system performance. The results shows that a potential fuel economy improvement around 10% with modern refrigerants and advancements in expander technology.

To increase the efficiency of ORC, the most suitable working fluids are recommended. Tian *et al.* [2] has studied the fluids and parameters optimization for ORCs used in the exhaust heat recovery of the Internal Combustion Engine (ICE). From the study, it has been demonstrated that R141b, R123 and R245fa present the highest ηth and Pnet values ranging from 16.6% to 13.3% (ηth value), and from 60 to 49 kJ/kg (Pnet value). Meanwhile, the three fluids also correspond to the lowest Epc values ranging from 0.30 to 0.35$/kWh and lowest A/Wnet values ranging from 0.436 to 0.516 m2/kW. The optimum evaporating pressures for R141b, R123, and R245fa are ranging from 2.8 MPa to 3.6 MPa. Zhu *et al.* [3] have performed a thermodynamic analysis in order to select appropriate working fluids and optimized parameters for medium-temperature geothermally-powered ORC. In this paper, R245fa is mixed with R601a at geothermal water temperature of 110°C. Based on the thermodynamic analysis, the results show that the zeotropic mixture R245fa/R601a (0.4/0.6) has the highest performance. When the evaporating temperature reaches 67°C, the outlet temperature of geothermal water is 61°C, the net power output is the highest and the thermal efficiency is about 9%.

Parametric optimization is the main concern in ORC in order to achieve the higher efficiency of the cycle. Kai *et al.* [4] have conducted a parametric optimization of a low temperature ORC system. To conduct the optimization, net power output per unit mass of geothermal water is set to be the objective function in Genetic Algorithm. From the optimization, under the assumption condition that the optimal level constitution of system parameters is determined as the working fluid of R227ea, it demonstrates that the power output of the system of 596 kW and the corresponding power output per unit mass of geothermal water of 11.91 kJ/kg. The effect of pinch temperature in evaporator on the power output per unit mass is more seriously than that of superheating. It is about 7 times in terms of increasing 1K of R227ea. A thermo-economic multi-objective optimization of ORC for exhaust waste heat recovery of a diesel engine has been performed by Yang *et al.* [5]. The effects of the four key parameters that include evaporation pressure, superheat degree, condensation temperature and exhaust temperature at the outlet of the evaporator on the thermodynamic performance and economic indicators of the ORC system are investigated. Six different working fluids are used to conduct the optimization and R245fa has been considered as the most suitable working fluid for ORC waste heat application of the diesel engine with consideration of thermoeconomic performances, environmental impacts and safety levels. To determine the most suitable working fluid for power generation from a low temperature geothermal heat source, Caceres *et al.* [6] have performed the optimization of an ORC. For each working fluid, an optimal configuration is obtained by conducting the optimization of the second law efficiency. The result shows that propylene and R1234yf are the best working fluid with a critical temperature close to the maximum temperature of the cycle that gives the highest plant efficiencies. Yang *et al.* [7] have conducted the parametric optimization and performance analysis of the ORC system by using Genetic Algorithm to recover the exhaust waste heat of a diesel engine. Three key parameters: evaporation pressure, superheat degree and condensation temperature, are analysed to study their effects on the system performance with POPA (net power output per unit heat transfer area) and also EDR (exergy destruction rate) are set as objective functions. From the analysis, at engine maximum rated power point, the ORC system can achieve maximum POPA of 0.74 kW/m^2, and the ratio of maximum effective heat transfer area to the actual area of the evaporator is 69.19%.

This paper presents the thermodynamic analysis and performance optimization of an ORC as waste heat recovery system on an aircraft engine. The primary objective is to investigate the effect of the parameters design on engine performance, including Thrust Specific Fuel Consumption (TSFC) and Specific Thrust of the engine.

2. Thermodynamic model of an ORC system

An ORC consists of the same principle as ideal Rankine cycle with a different in the type of working fluid. ORC is a closed-loop system where the working fluid circulates through four main components: an evaporator, a pump, a turbine and a condenser. In this study, a regenerator is integrated to the ORC cycle to increase the thermal efficiency of the system. The ORC system is integrated in the middle of the low pressure turbine exit duct and core nozzle of an aircraft engine. Figure 1 shows the schematic diagram of a basic ORC system with regenerator integrated to a jet engine of an aircraft. ORC system is composed of an evaporator, a condenser, a turbine and a working fluid pump and a regenerator with a specific function to preheat the working fluid before entering the evaporator. This function is being considered in order to increase the thermal efficiency of the system. The system described in this study is integrated to the jet engine in the middle of the low pressure turbine exit duct and the core nozzle.

Figure 1 illustrates the connection between the Brayton cycle and the ORC cycle, which is located at the evaporator. The system begins to operate at the outlet of the cold side of the regenerator, which is at station 1 in Figure 1, with a specific heat transfer from the exhaust jet engine and a specific evaporator outlet temperature. The inlet temperature of the working fluid flow to the evaporator is estimated by the formula described in the analysis section and the system is analyzed to obtain the parametric optimization. Figure 2 shows the T-s diagram of the cycle involved in the system. The T-s diagram described is of the ORC system with regenerator integrated to a jet engine.

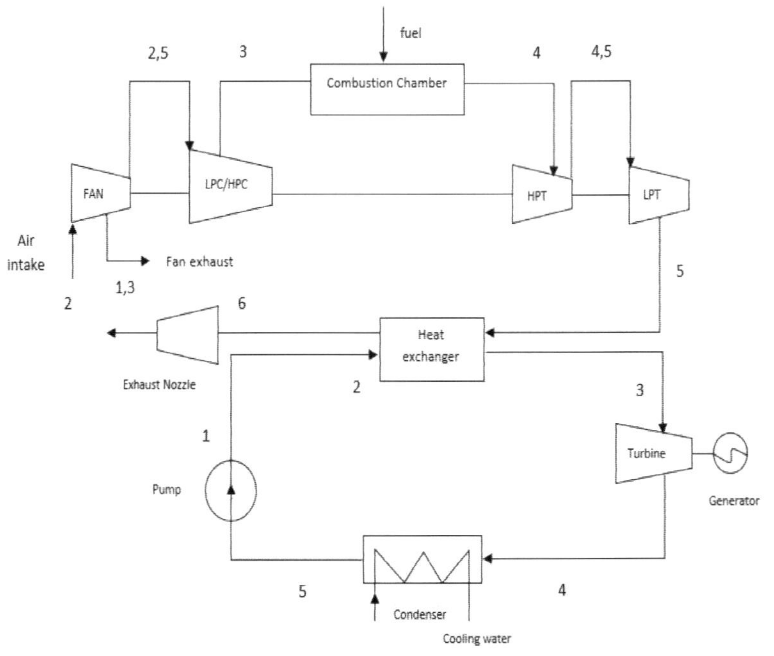

Figure 1. Schematic diagram of the Brayton-Rankine combined cycle

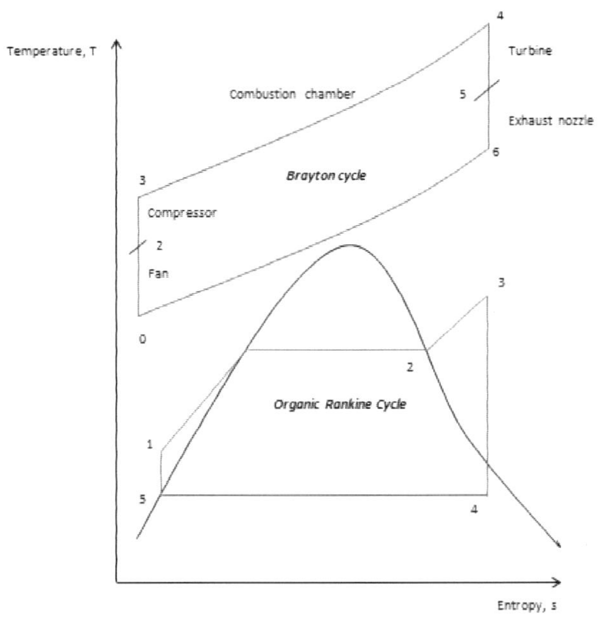

Figure 2. T-s diagram of the Brayton-Rankine combined cycle

The type of evaporator used in this study is the shell and tube heat exchanger, with one pass tube-side. Shell and tube heat exchanger consist of a series of tubes mounted inside a cylindrical shell. One set of these tubes contains the fluid that must be either heated or cooled. Then, the second fluid runs over the tubes that are being heated or cooled such that it can either provide the heat or absorb the heat required. From the counter flows, the heat is being exchanged between the two different fluids with different temperatures. In this study, several assumptions for the thermodynamic model are listed as follows:

1. The ORC system operates under steady state conditions.
2. There is no pressure drop in the pipes and the components.
3. The heat losses in each component are also neglected.
4. The isentropic efficiencies of the pump and the expander are 0.7 and 0.87 respectively.
5. The heat transfers to or from the surrounding environment are neglected.

Table 1. Main parameters of ORC system with R245fa as the working fluid for the turbofan engine

Description	Unit	Value
Mass flow rate of R245fa	kg/s	3.837
Exhaust heat temperature	K	720 – 1040
Turbine inlet pressure	Mpa	2.2
Turbine inlet temperature	K	392
Inlet temperature of R245fa	K	282
Outlet temperature of R245fa	K	393
Turbine efficiency	-	0.87
Pump efficiency	-	0.7
Estimated required surface area of evaporator	m^2	23.72

A parametric optimization of Turbine Inlet Temperature (TIT) and the preheating temperature of the working fluid is performed to obtain the minimum value of TSFC along the saturated vapor line in sub-critical region for R245fa working fluids. In this study, TSFC and Specific Thrust are evaluated as objective functions with TIT and preheating temperature of the working fluid are chosen as the main variables. The genetic algorithm optimization functions can be described as:

Max (Specific Thrust) = *f (TIT, preheating temperature)*
Min (TSFC) = *f (TIT, preheating temperature)*

Table 2: The lower and upper bounds of decision variables

Decision variable	**Upper bound**	**Lower bound**
Turbine inlet temperature (K)	350	450
Preheating temperature (K)	250	350

TIT can be defined as the temperature of the air gas mixture at the inlet of the gas turbine and it is one of the most critical parameters that influences the gas turbine performance. For a better gas turbine performance, it is desirable to have a higher turbine inlet temperature. Both the power output and the thermal efficiency can be improved by increasing the TIT.

3. Result and discussion

Figure 3 and 4 show the variation of TSFC and Specific Thrust of the engine. The range of TIT set for this system is between 350 K – 450 K whereas for the preheating temperature, it is set between 250 K – 350 K. It is obvious that, as the TIT and preheating temperature increases, the Specific Thrust of the engine increases linearly with decrease in TSFC. The optimization of TIT and preheating temperature provide a huge impact on the performance of the engine. Thus, by performing parametric optimization on the engine system, the performance of the engine can be improved.

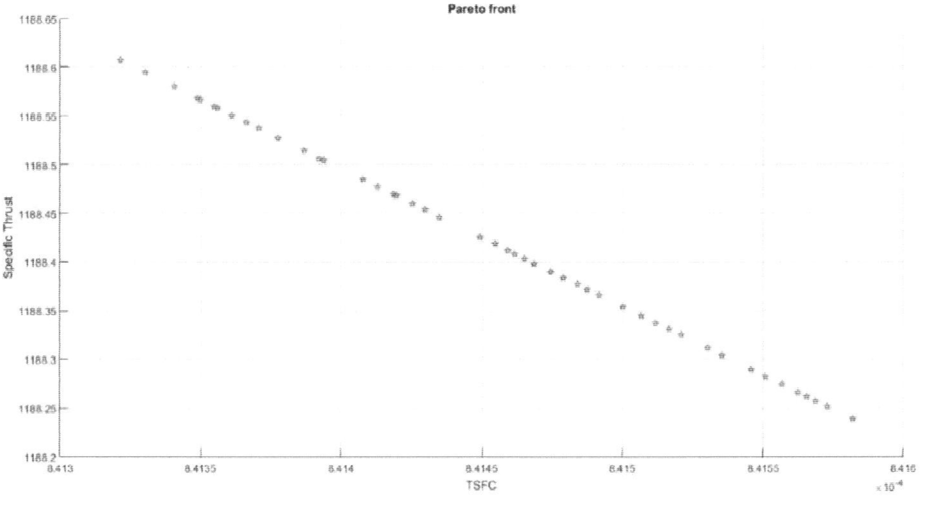

Figure 3. Optimization of TIT on the TSFC and Specific Thrust

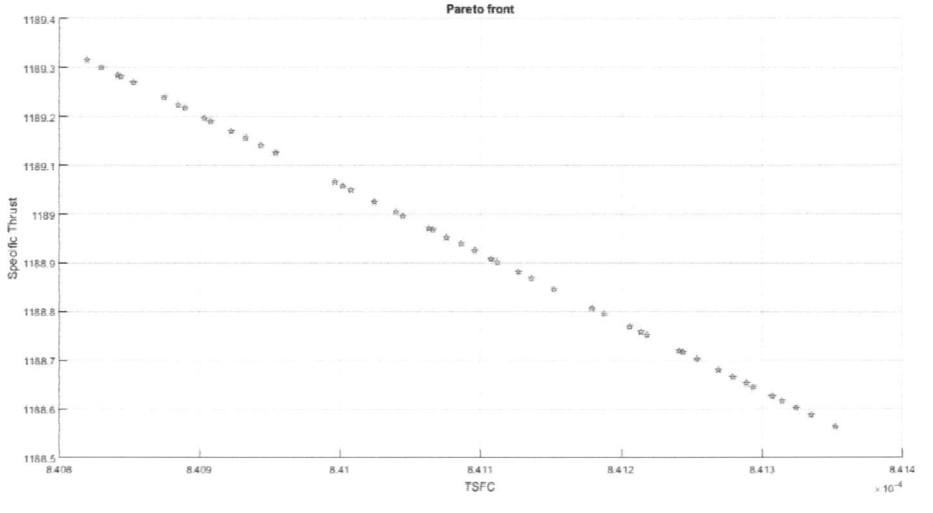

Figure 4. Optimization of preheating temperature on the TSFC and Specific Thrust

4. Conclusion

In this study, thermodynamic analysis and parametric optimization of the ORC Waste Heat Recovery for turbofan engine are conducted. Two-spool mixed flow turbofan engine is used in this study to investigate the beneficial of using ORC as waste heat recovery method in the engine. The working fluid used in this study is R245fa. Based on the first and second law of thermodynamics, the turbine inlet temperature and preheating temperature of the working fluid are the parameters that can affect the performance of the system. By using Genetic Algorithm, parametric optimization can be conducted. The optimization of TIT and preheating temperature is performed and it shows that increases in TIT and preheating temperature of the working fluid will increase the net power output of the system. The presented research proves that ORC cycle for waste heat recovery provides potential advantages when integrated to an aircraft engine. It is significant that, by using ORC as heat recovery, the minimization of TSFC and maximization of specific thrust of a turbofan engine can be achieved.

References

[1] Alshammari F, Karvountzis-Kontakiotis A, Pesiridis A, Minton T. Radial expander design for an engine Organic Rankine Cycle waste heat recovery system. Energy Procedia 2017; 129: 285–292

[2] Tian H, Shu G, Wei H, Liang X, Liu L. Fluids and parameters optimization for the organic Rankine cycles (ORCs) used in exhaust heat recovery of Internal Combustion Engine (ICE). Energy 2012; 47(1): 125–136

[3] Zhu J, Kang Z, An Q, Li T. Parametric optimization of organic Rankine cycle with R245fa/ R601a as working fluid. Transactions of Tianjin University 2015; 21: 69-75

[4] Kai Z, Mi Z, Yabo W, Zhili S, Shengchun L, Jinghong N. Parametric optimization of low temperature ORC system. Energy Procedia 2015; 75: 1596–1602.

[5] Yang F, Zhang H, Song S, Bei C, Wang H, Wang E. Thermoeconomic multi-objective optimization of an organic Rankine cycle for exhaust waste heat recovery of a diesel engine. Energy 2015; 93: 2208–2228

[6] Cáceres I, Agromayor R, Nord L O. Thermodynamic optimization of an organic rankine cycle for power generation from a low temperature geothermal heat source. 58th SIMS, 2017

[7] Yang F, Zhang H, Bei C, Song S, Wang E. Parametric optimization and performance analysis of ORC (organic Rankine cycle) for diesel engine waste heat recovery with a fin-and-tube evaporator. Energy 2015; 91: 128–141

IOP Conf. Series: Materials Science and Engineering **405** (2018) 012011 doi:10.1088/1757-899X/405/1/012011

An overview of horizontal-axis Magnus wind turbines

O F Marzuki, A S Mohd Rafie*, F I Romli, K A Ahmad and M F Abdul Hamid

Department of Aerospace Engineering, Faculty of Engineering,
Universiti Putra Malaysia, 43400 Serdang, Selangor, Malaysia.

*shakrine@upm.edu.my

Abstract. Magnus force shows extraordinary results in producing lift force, hence gives rise to its application. One of the applications is by replacing airfoil-shaped blades of wind turbine with the rotating cylinder blades. This is known as Magnus Wind Turbine (MWT), where it generates lift force on the rotating cylinder that perpendicular to the incoming wind flows. Due to high consumption of fossil fuel, natural gas and coal that risk our and earth health. Thus to overcome this conundrum, MWT is one of the innovative approaches to harvest the low wind speed energy. This paper presents an overview of the horizontal-axis MWT state-of-the-art.

1. Introduction

Sustainable energy is a necessity to meet the current and future generation needs of energy. The basic principle of harvesting wind energy is by converting the incoming wind flow to kinetic energy through mechanical mechanism and using a generator to generate electricity [1–3]. As one of natural resources, wind energy can be used repeatedly and will never be depleted. However, availability of wind energy resources is dependent on the geographical location and topography [4–5]. Nowadays, harvesting high speed wind energy is mainly dominated by horizontal axis wind turbine (HAWT) with airfoil-shaped blades. The airfoil-shaped-blades wind turbine requires a minimum wind speed of 4 meters per second (m/s) to generate minimal power and wind speed of 16 m/s to generate optimum power [6]. Due to this limitation, an innovative type of wind turbines is researched and designed for harvesting wind energy at lower wind speeds that can produce optimum power.

One of the types of wind turbines that can harvest wind energy at lower wind speed condition is the horizontal-axis Magnus Wind turbines (MWT) [7]. The main difference of MWT in comparison to the wind turbine utilizing airfoil-shaped blades lies in the means of extracting wind energy. By utilizing rotating cylinder blades, MWT can harvest wind energy at low wind speed condition. Additionally, the performance of MWT can be increased by enhancing the surface of the rotating cylinder blades. There are several ways that researchers have innovated the MWT rotating cylinder blades performance such as by utilizing dimple and fin [8–9]. Moreover, the use of sandpaper to add surface roughness on the rotating cylinder blades can further improve the torque generated by the MWT rotor [10–12]. Figure 1 shows the MWT with spiral fins coiled around the rotating cylinder blades that has been developed in Japan.

This paper offers an overview of the state-of-the-arts for horizontal-axis MWT that can be used for reference in developing low wind speed wind energy harvester. This paper also aims to uncover some of operational principles surrounding MWT such as with regard to the number of blades and types of surface roughness on the rotating cylinders.

IOP Conf. Series: Materials Science and Engineering **405** (2018) 012011 doi:10.1088/1757-899X/405/1/012011

Figure 1. Magnus wind turbine with fin-like on rotating cylinder blades [13]

2. Magnus wind turbine

Commonly, a wind turbine with airfoil-shaped blades utilizes and prioritizes a smooth surface blades to produce high lift and optimum power at high wind speed condition. However, the lift performance drops considerably when there are some irregularities to the surface of the airfoil-shaped blades [14]. In addition, high noise level is produced by the wind turbine due to the rotor rotating at a high speed [15]. Hence, the application of MWT as an alternative of the wind turbine using airfoil-shaped blades is taken into deliberation. As the MWT uses rotating cylinder as blade, it can generate high torque and lift at low wind speed condition by further increasing the cylinder blades rotation speed as compared to the airfoil-shaped blades that is limited by the angle of attack. Besides, in comparison with the wind turbine with airfoil-shaped blades, the MWT can also reduce environmental hazard and noise due to a slower rotor rotation [7].

To overcome the drawback of the wind turbines with airfoil-shaped blades, Bychkov *et al.* [7] have demonstrated an experiment on the MWT using 6 smooth rotating cylinders blades with endplates as shown in Figure 2. They have concluded that 6 rotating cylinders blades and the cylinder aspect ratio of 15 are the most optimized for MWT [16]. Moreover, they have proposed that MWT will produce higher performance in low wind speed condition than the wind turbine with airfoil-shaped blades. On the other hand, based on Bychkov *et al.* [7], MWT rotating cylinder blades are designed by using computational fluid dynamics while Mara [17] has offered the rotating cylinder blades design profile characteristics for MWT blades.

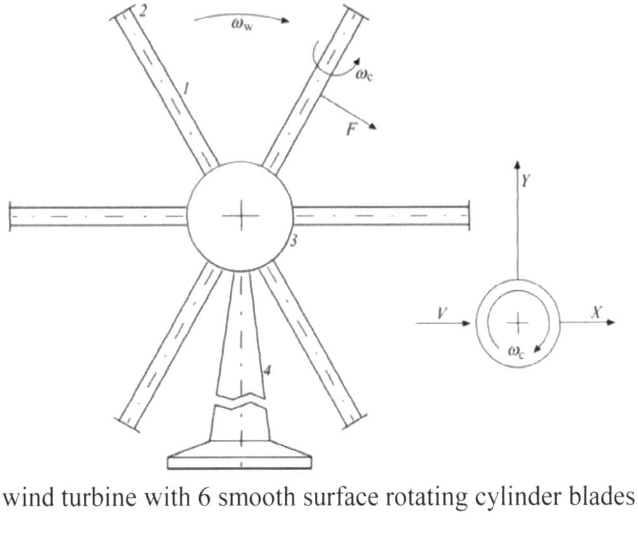

Figure 2. Magnus wind turbine with 6 smooth surface rotating cylinder blades and endplates [7]

Murakami and Ito [18] have patented a MWT with 6 spiral fins rotating cylinder blades known as Spiral MWT. Afterwards, they have upgraded the patent for the Spiral MWT with 5 spiral fins rotating cylinder blades [9]. They have demonstrated that the rotating cylinder blades with spiral fins coiled around the cylinder can produce more than 2 times higher lift coefficient and torque as compared to the smooth surface rotating cylinder blades. Figure 3 depicts the Spiral MWT patents from 6 rotating cylinder blades to 5 rotating cylinder blades. Later on, Kato *et al.* [19] and Murakami *et al.* [20] have conducted Particle Image Velocimetry (PIV) experiment to investigate the effectiveness of MWT that uses spiral fins coiled around the rotating cylinder blades.

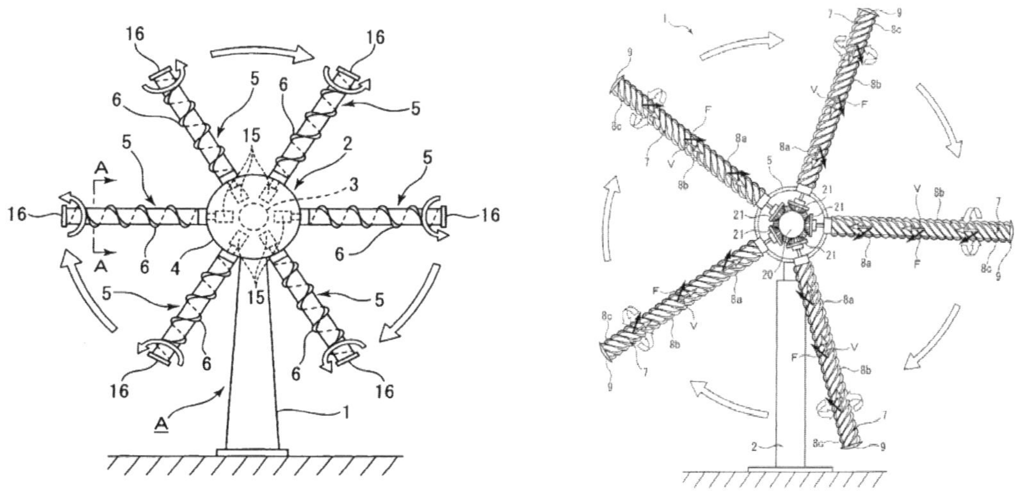

Figure 3. Spiral MWT patents upgraded from 6 to 5 rotating cylinder blades [9,18]

Based on prior patents of MWT using spiral fins [18], the researchers have fabricated MWT model with 6 spiral fins coiled around cylinder blades to find the optimize value of power generated based on the cylinder blades speed rotation [21–23] and proposed an algorithm to control the rotating cylinder speed [8]. Moreover, Corrêaet *et al.* [24] have proposed a performance emulator for MWT. Figure 4 shows the MWT model utilized in their performance optimization experiment with an oversized rotor hub.

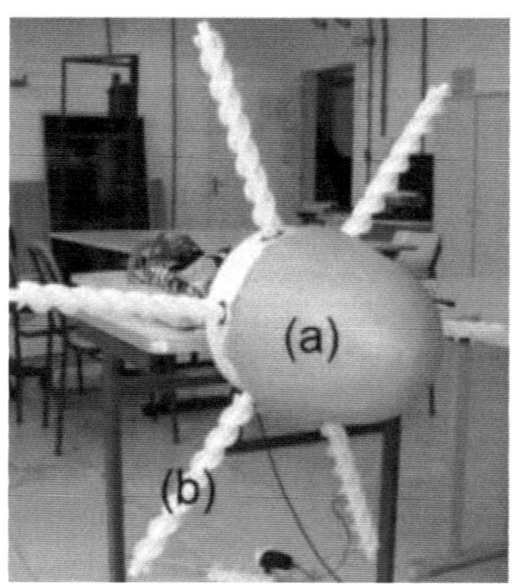

Figure 4. MWT model with spiral fins coiled around cylinder blades [22]

Giudice and La Rosa [25] have presented a MWT prototype that is also developed for harvesting water stream energy known as Chiral Rotor. This MWT prototype utilizes 4 smooth surface rotating cylinder blades and it is designed to harvest energy from water channels like drainage and irrigation. They have concluded that the Chiral Rotor prototype has a great prospect toward hydroelectricity as an alternative to wind energy harvester. Figure 5 shows the Chiral Rotor prototype with 4 short rotating cylinder blades. Moreover, they emphasize that, the Chiral Rotor prototype use smooth surface on the cylinder blades due to the absence of knowledge and research regarding the surface roughness effect on rotating cylinder blades. Meanwhile, Ito *et al.* [26] have investigated a prospect of 3 cylinder blades shaped like Savonius wind turbine with endplates and concluded it as one of the possible combinations as shown in Figure 6.

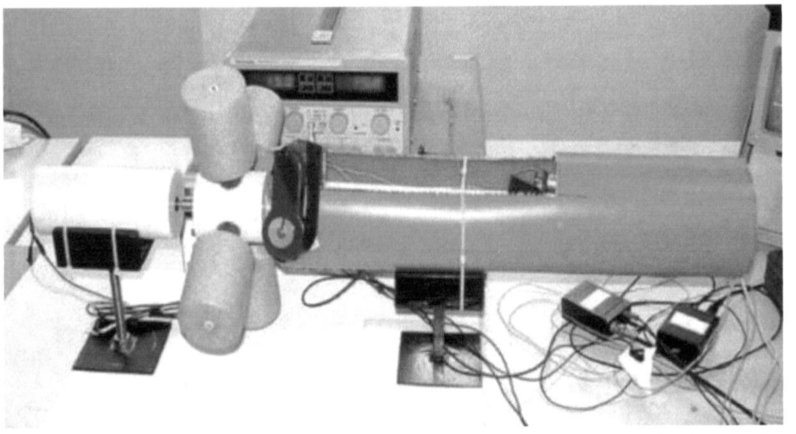

Figure 5. Chiral rotor prototype with 4 smooth surface rotating cylinder blades [25]

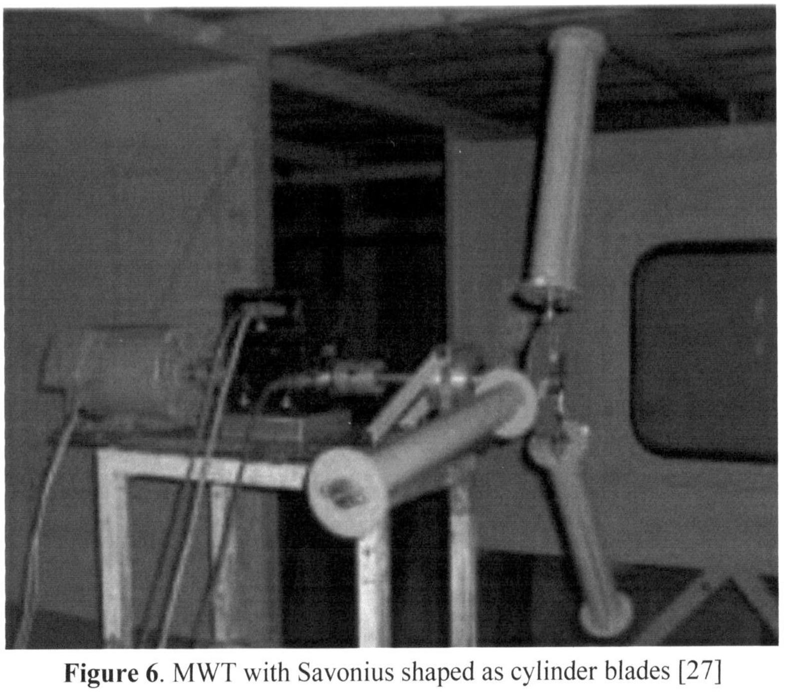

Figure 6. MWT with Savonius shaped as cylinder blades [27]

Marzuki *et al.* [10, 28, 29] have investigated the torque performance impact of sandpaper surface roughness types on MWT with 5 rotating cylinder blades. The finding highlights that, as the surface

IOP Conf. Series: Materials Science and Engineering **405** (2018) 012011 doi:10.1088/1757-899X/405/1/012011

roughness is changed from smooth to rough sandpaper, the torque produced by the MWT significantly increases up to 5 times compared to that with smooth surface [10]. Figure 7 shows the MWT model without and with sandpaper enhancement inside wind tunnel. Meanwhile, researchers Sakipova *et al.* [12, 30] have investigated the performance of MWT with porous surface roughness with 3 cylinder blades and later on with 2 cylinder blades as shown in Figure 8. In an analysis of the effect of porosity surface on the rotating cylinder, Kussaiynov *et al.* [31] have found that it gives a notable improvement of lift force around 1.5 times in comparison to the smooth surface cylinder.

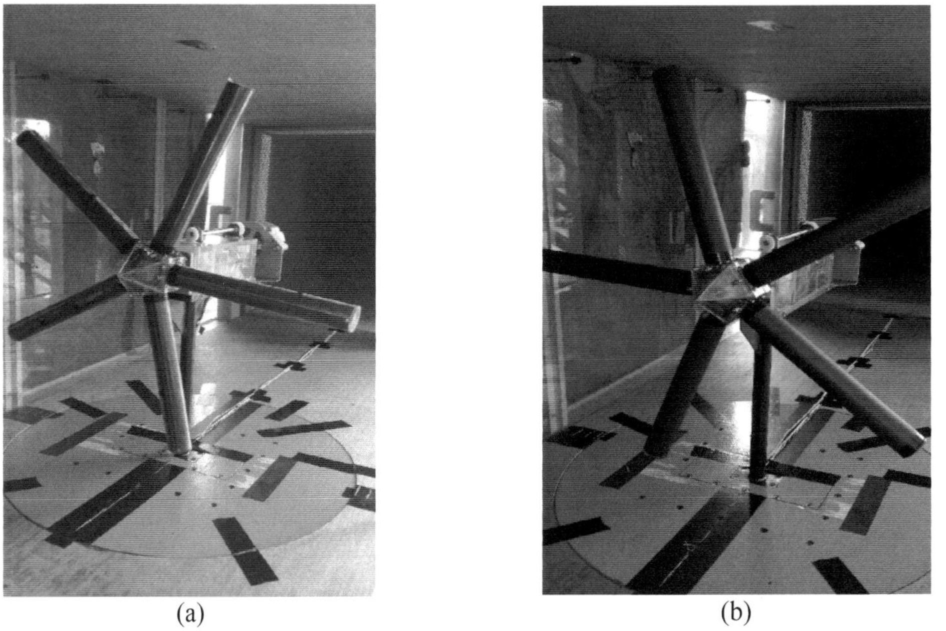

(a) (b)

Figure 7. MWT model with (a) sandpaper attachment, (b) smooth [10]

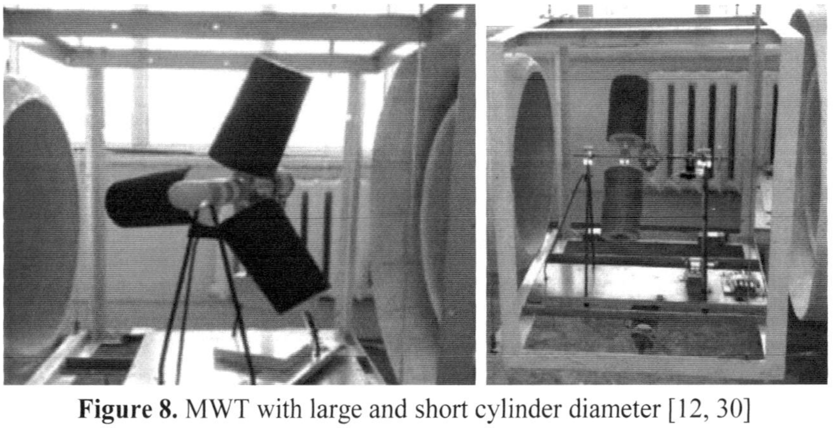

Figure 8. MWT with large and short cylinder diameter [12, 30]

There are various researchers that have proposed numerical theories and run analytical analysis on MWT [32–37]. In additional, Ogretim *et al.* [38] have proposed solar enhanced MWT by covering the rotating cylinder surface with solar cells. However, with the absence of MWT experimental approach, restraints must be made when implementing the proposed theories. Sedaghat [33] has reviewed and concluded that, as more experimental study on MWT are being conducted, it will have bright prospect in harvesting wind energy.

3. Conclusion

Overall, this overview strengthens the idea that horizontal-axis MWT is feasible as sustainable energy harvester. There are several ways to improve the MWT performance and one of them is by utilizing the surface roughness effect on rotating cylinder that include fins, dimples, groove, porous, sandpapers and so on. Moreover, considering all evidences, it seems that there is no fixed number of cylinder blades for MWT. Despite its exploratory nature, this overview offers some insights into current state-of-the-art horizontal-axis MWTs. Further research could also be conducted to determine the power usage and the power produced from the MWT, and the effect of number of rotating cylinder blades. A reasonable approach to tackle this issue could be to do an experimental research.

Acknowledgement
The authors acknowledge the financial support for this study from the Ministry of Higher Education, Malaysia through Exploratory Research Grant Scheme – ERGS/1/11/TK/UPM/02/12

References

[1] Daut I, Razliana R N, Irwan Y M, Farhana Z. A study on the wind as renewable energy in Perlis, northern Malaysia. Energy Procedia 2012; 18: 1428–33
[2] Haidar A M, John P N, Shawal M. Optimal configuration assessment of renewable energy in Malaysia. Renew. Energy 2011; 36: 881–8
[3] Qian X, He Y. Wind power turbine and its aerodynamic characteristics. Int. J. Educ. Manag. Eng. 2012; 2: 80–7
[4] Goh H H, Lee S W, Chua Q S, Goh K C, Teo K T K. Wind energy assessment considering wind speed correlation in Malaysia. Renew. Sustain. Energy Rev. 2016; 54: 1389–400
[5] Khatib T, Sopian K, Mohamed A, Ibrahim M Z. Sizing of a wind charger at minimum cost for remote housing electrification: A case study for nine coastal sites in Malaysia. Energy Build. 2012; 51: 185–90
[6] Balat M. A review of modern wind turbine technology. Energy Sources, Part A Recover. Util. Environ. Eff. 2009; 31: 1561–72
[7] Bychkov N M, Dovgal A V, Kozlov V V. Magnus wind turbines as an alternative to the blade ones. J. Phys. Conf. Ser. 2007; 75: 12004
[8] Jinbo M, Ceretta Moreira M, Lellis Hoss D, Farret F A, Cardoso Junior G. Fixed and adaptive step HCC algorithms for MPPT of the cylinders of Magnus wind turbines. 3rd Renewable Power Generation Conference, 2014
[9] Murakami A N. Magnus Type Wind Power Generator 1, 2010
[10] Marzuki O F, Mohd Rafie A S, Romli F I, Ahmad K A. Magnus wind turbine: the effect of sandpaper surface roughness on cylinder blades. Acta Mech. 2018; 229: 71–85
[11] Marzuki O F, Rafie A M, Romli F I, Ahmad K A. An experimental investigation on the effect of surface roughness on the performance of Magnus wind turbine. ARPN J. Eng. Appl. Sci. 2015; 10
[12] Sakipova S E, Tanasheva N K, Kussaiynova A K. Study of aerodynamics of a two-bladed wind turbine with porous-surfaced cylindrical blades. Eurasian Phys. Tech. J. 2017; 14: 120–4
[13] MECARO Co. Ltd. Spiral Magnus: Latest evolution of wind turbine technology, 2007
[14] Soltani M, Birjandi H, Moorani M. Effect of surface contamination on the performance of a section of a wind turbine blade. Sci. Iran. 2011; 18: 349–57
[15] Dai K, Bergot A, Liang C, Xiang W-N, Huang Z. Environmental issues associated with wind energy – A review. Renew. Energy 2015; 75: 911–21
[16] Bychkov N M, Dovgal A V, Sorokin A M. Parametric optimization of the Magnus wind turbine. Int. Conf. Methods Aerophysical Res., 2008
[17] Mara B K, Mercado B C, Mercado L A, Pascual J M, Lopez N S. Development and validation of a CFD model using ANSYS CFX for aerodynamics simulation of Magnus wind rotor blades. International Conference on Humanoid, Nanotechnology, Information Technology,

Communication and Control, Environment and Management (HNICEM), 2014
[18] Murakami N and Ito J. Magnus Type Wind Power Generator U.S. Pat. No. 7,504,740 B2, 2009
[19] Kato H, Koike S, Nakakita K, Ito T, Shiohara T, Kato C, Iida A, Doi Y, Kato Y, Miura Y. PIV measurements of unsteady flowfields around Magnus wind turbines with spiral fins. 10th Int. Symp. Part. IMAGE Velocim. – PIV13, 2013
[20] Murakami N, Hasegawa H, Haniu T, Nakamura M. The mechanism of flow field around a rotating cylinder with fins for high performance Magnus wind turbine. J. Energy Power Eng. 2016; 10: 23–31
[21] Correa L C, Lenz J M, Ribeiro C G, Trapp J G, Farret F A. MPPT for Magnus wind turbines based on cylinders rotation speed. Brazilian Power Electronics Conference (IEEE), 2013
[22] Jinbo M, Farret F A, Cardoso Junior G, Dreher J R, Lellis Hoss D. MPPT for Magnus wind system with DC servo drive for the cylinders and boost converter. 3rd Renewable Power Generation Conference, 2014
[23] Jinbo M, Farret F A, Cardoso Junior G, Senter D, Lorensetti M F. MPPT for Magnus wind system with DC servo drive for the cylinders and boost converter. J. Wind Energy 2015; 148680
[24] Corrêa L C, Lenz J M, Ribeiro C G, Farret F A. Magnus wind turbine emulator with MPPT by cylinder rotation control. J. Dyn. Syst. Meas. Control 2018; 140: 101012–7
[25] Giudice F, La Rosa G. Design, prototyping and experimental testing of a chiral blade system for hydroelectric microgeneration. Mech. Mach. Theory 2009; 44: 1463–84
[26] Ito A, Nishizawa Y, Ushiyama I. An experiment study on the Savonus type Magnus wind turbine. Japan Soc. Mech. Eng. 2009
[27] Ito A, Kawashima S, Nishizawa Y, Ushiyama I, Komatinovic N. A Study on Savonius type Magnus wind turbine. Eur. Wind Energy Conf. Exhib., 2007
[28] Marzuki O F, Mohd Rafie A S, Romli F I, Ahmad K A. An experimental investigation on the effect of surface roughness on the performance of Magnus wind turbine. ARPN J. Eng. Appl. Sci. 2015; 10
[29] Marzuki O F, Rafie A S M, Romli F I, Ahmad K A. Torque performance study of Magnus wind turbine. Int. Rev. Mech. Eng. 2015; 9: 38–42
[30] Sakipova S E, Tanasheva N K, Kivrin V I, Kussaiynova A K. Study of aerodynamic characteristics of the model of a wind turbine with rotating cylinders. Eurasian Phys. Tech. J. 2016; 1–6
[31] Kussaiynov K, Tanasheva N K, Turgunov M, Shaimerdenova G M, Alibekova A R. The effect of porosity on the aerodynamic characteristics of a rotating cylinder. Mod. Appl. Sci. 2015; 9: 215–22
[32] Luo D, Huang D, Wu G. Analytical solution on Magnus wind turbine power performance based on the blade element momentum theory. J. Renew. Sustain. Energy 2011; 3
[33] Sedaghat A. Magnus type wind turbines: Prospectus and challenges in design and modelling. Renew. Energy 2014; 62: 619–28
[34] Sedaghat A, Samani I, Ahmadi-Baloutaki M, El Haj Assad M, Gaith M. Computational study on novel circulating aerofoils for use in Magnus wind turbine blades. Energy 2015; 91: 393–403
[35] Sun X, Zhuang Y, Cao Y, Huang D, Wu G. A three-dimensional numerical study of the Magnus wind turbine with different blade shapes J. Renew. Sustain. Energy 2012; 4: 63139
[36] Richmond-Navarro G, Calderon-Munoz W R, LeBoeuf R, Castillo P. A Magnus wind turbine power model based on direct solutions using the blade element momentum theory and symbolic regression. IEEE Trans. Sustain. Energy 2017; 8: 425–30
[37] Gustavo R N, Noel U S, Giancarlo R. High correlation models for small scale Magnus wind turbines. 5th International Conference on Renewable Energy: Generation and Applications (ICREGA), 2018
[38] Ogretim E O, Uygun D, Koklu M O. Analytical evaluation of solar enhanced Magnus effect wind turbine concept. Int. J. Renew. Energy Res. 2016; 6: 2–7

IOP Conf. Series: Materials Science and Engineering **405** (2018) 012012 doi:10.1088/1757-899X/405/1/012012

Ballistic impact analysis of double-layered metal plates

Rizal Zahari[1,2*], John Regan Pillai[1], Andrew Ordys[1], Mohamed Thariq Hameed Sultan[2] and Noorfaizal Yidris[2]

[1]Systems Engineering Department, Military Technological College, P.O Box 262 PC 111, Muscat, Sultanate of Oman
[2]Aerospace Manufacturing Research Centre, Department of Aerospace Engineering, University Putra Malaysia, 43400 Serdang, Selangor, Malaysia

*Rizal.Zahari@mtc.edu.om

Abstract. In the present study, ballistic impact analysis of double layered circular metal plates with different arrangements were investigated numerically using the finite element method. Nine different double layered plates configuration consist of steel 4340 (st), Aluminium 2014-T3 (Al), or titanium Ti-6A1-4V (Ti) having the same thickness were considered in this study and the energy absorption capability of each configuration were determined and compared. In order to validate the results obtained via the numerical method, an experimental ballistic test utilising a single stage gas gun was performed on single layer Al plates using a blunt projectile. It was found that the results obtained from numerical analysis agree well with the experimental results, which confirmed the accuracy of the numerical analysis procedure. The non-linear explicit finite element analysis employing Johnson-Cook plasticity material model coupled with was carried out using ABAQUS commercial package. A parametric study was conducted to determine the best configuration in terms of the energy absorption capability under ballistic impact. It was observed from the analysis that the double layered plate with titanium (front plate facing the projectile) and steel 4340 (back plate) produced the highest energy absorption with 279.72 J. However, the highest specific energy absorption (SEA) was given by titanium (front) and Aluminium (back) plate with 2830.45 J/kg.

1. Introduction

The design and development of armour plates for protection against ballistic impact has long been of interest in military and civilian applications [1]. A multi-layer configuration that consisted of several parallel plates is proposed to replace the design using monolithic or single plates. When a single plate is replaced by several layered thin plates, the order, thickness and the number of layers will affect the failure modes, which leads to difference of ballistic resistance between various target configurations [2]. Although there were a number of research dealing with the ballistic behaviour of multi-layered plates [3-9], their scope was limited when compared to studies of monolithic plates. Moreover, the studies conducted on multi-layered armour plates using numerical analysis is also scarce [10-17]. Thus the study of multi-layered plates remains an open research topic since the conclusive results of their performance have not been obtained to date.

In this research, ballistic impact analysis of double layered circular metal plates with different order were investigated numerically using finite element method using nine different double layered plates configuration consisted of steel 4340 (st), Aluminium 2014-T3 (Al) or titanium Ti-6A1-4V (Ti) [18].

The thickness of each plate is assumed to be the same and the plates are assumed to be in contact with each other. The energy absorption capability of each configuration was determined and compared in order to determine the most optimum configuration.

2. Methodology

The current research is conducted based on tasks and stages as shown in Figure 1. The work is started with the fabrication of the Al 2024-T3 single plates, whose dimensions are in accordance to the NIJ standard requirement. A parametric study is carried out on nine different configurations for the double layered circular plates (50mm diameter and 3mm thickness for each layer) using the specific notation shown in Table 1. Finally, the optimum configuration based on specific energy absorbed is determined using the numerical analysis.

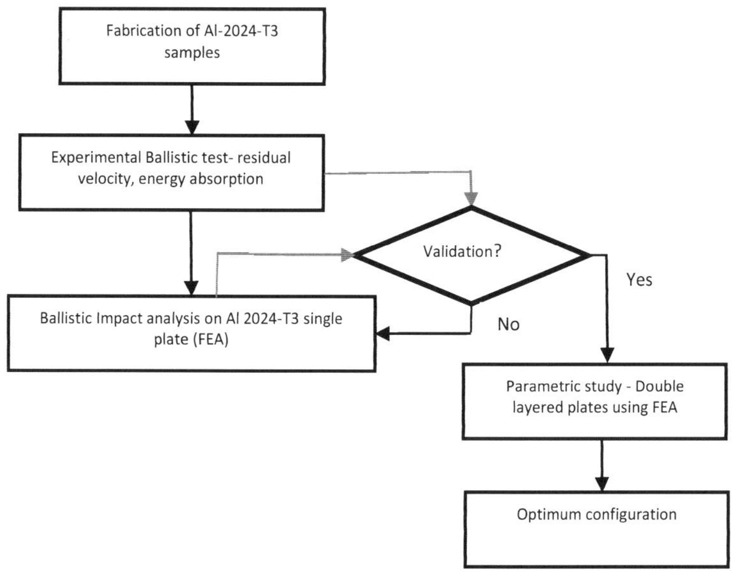

Figure 1. Flowchart for the overall research work

Table 1. Configuration for the double layered plates

Material (Front plate-Back plate)	Configuration
(Steel-steel)	*C1*
(Aluminium-aluminium)	*C2*
(Titanium-titanium)	*C3*
(Steel-titanium)	*C4*
(Titanium-aluminium)	*C5*
(Steel-aluminium)	*C6*
(Titanium-steel)	*C7*
(Aluminium-titanium)	*C8*
(Aluminium-steel)	*C9*

2.1. *Experimental ballistic impact*

A single stage gas gun was used to conduct the experimental ballistic test in which the projectiles were launched at different pressures and velocities, shown in Figure. 2. The test setup was developed based on NIJ NIJ-018.01 standard [19], as shown in Figure 3. The impact energy absorption was determined using the principle of conservation of energy, i.e. the change in kinetic energy by the projectile before and after the impact is equivalent to the energy absorbed by the plate.

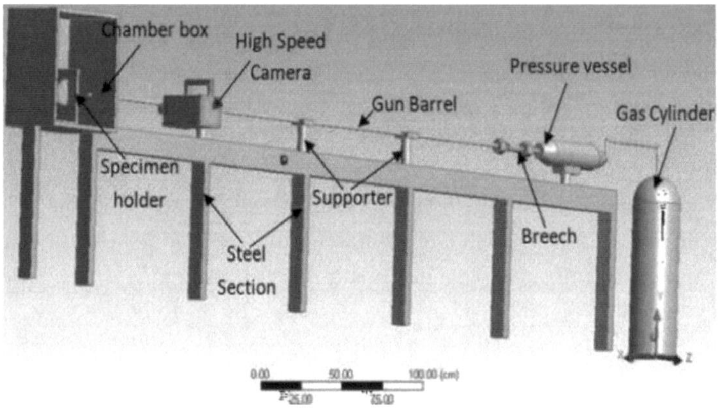

Figure 2. Gas gun tunnel instrument for impact test

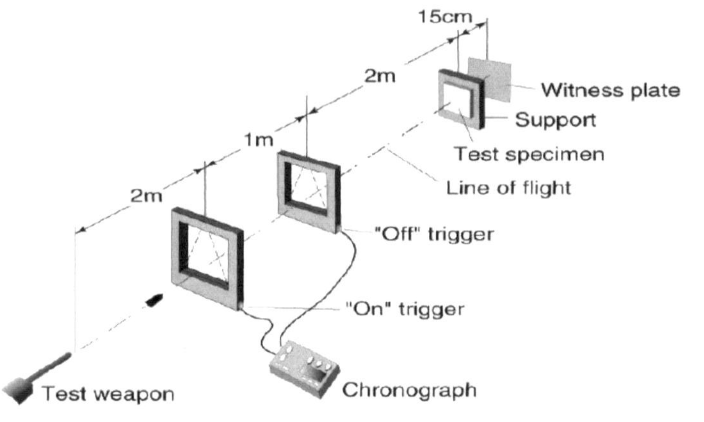

Figure 3. Ballistic test setup following the NIJ standard

Square targets with sides of 4 inches were fixed to a thick base plate by means of four bolts that were arranged in specimen holder in the target chamber box, as demonstrated in Figure 2. The test was done on a single square (100mm x 100mm x 3mm) monolithic aluminium 2024-T3 plate subjected to a blunt mild steel projectile with a diameter of 4.5 mm and 13.07mm length. A high speed camera was used to record the process of projectile penetrating target and the projectile motion. From the digital images, the travelled distance and projectile velocity before and after penetration were obtained. With these information, energy absorption by the plate can be estimated using the conservation of energy principle. The results obtained from this test served as a validation against the numerical results.

2.2. Numerical analysis

A numerical analysis using ABAQUS v6.13 finite element analysis software was used in the present study to investigate the ballistic performance of double layered plates having different arrangements as presented in previous Table 1. The two plates in each configuration are assumed to be in contact with each other using general contact formulation [20]. The non-linear dynamic explicit analysis has been employed in ABAQUS to accurately simulate the ballistic impact behaviour.

The Johnson & Cook plasticity formulation [21], which defines the flow stress as a function of the equivalent plastic strain, strain rate and temperature, was employed in the numerical simulations in order to predict damage behaviour in metal plates due to impacted projectile. The following relation in Eqn. 1 expresses the dynamic flow stress, where *A*, *B*, *m*, and *n* are material parameters, T, T_r, and T_m are temperature, room temperature and melting point, and $\dot{\varepsilon}_0$is a reference plastic strain rate.

$$\sigma_e = [A + B(\varepsilon_e^p)^n]\left[1 + C\ln\left(\frac{\dot{\varepsilon}_e^p}{\dot{\varepsilon}_0}\right)\right]\left[1 - \left(\frac{T - T_r}{T_m - T_r}\right)^m\right], \quad (1)$$

In the Johnson–Cook failure model [22], a damage parameter D is defined as in Eqn. 2,

$$D = \int \frac{1}{\varepsilon_f} d\varepsilon_e^p. \quad (2)$$

When D reaches 1, the failure occurs so that element will no longer withstand the tension and then will be deleted in simulations. Failure strain ε_f is defined in Eqn. 3, where d_1, d_2, d_3, d_4 and d_5 are material constants and P is hydrostatic pressure [22].

$$\varepsilon_f = \left(d_1 + d_2 e^{d_3 \frac{P}{\sigma_e}}\right)\left[1 + d_4 \ln\left(\frac{\dot{\varepsilon}_e^p}{\dot{\varepsilon}_0}\right)\right]\left[1 + d_5\left(\frac{T - T_r}{T_m - T_r}\right)\right], \quad (3)$$

Many studies show good agreements between predictions of Eqn. 3 and the experimental results of the steel and aluminium alloy at high strain rate [23, 24]. The yield surface parameters and failure strain parameters for Al 2024-T3, Ti-6A1-4V and Steel 4340 are given by Table 2 and Table 3, respectively. Table 4 presents the mass of each material and the projectile. These values are required to compute the energy absorb and the specific energy absorption of the double plates.

Table 2. Yield surface parameters Al 2024-T3, Ti-6A1-4V and Steel 4340 [17, 22]

Material	**Al 2024-T3**	**Ti-6A1-4V**	**Steel 4340**
Parameter	**Value**	**Value**	**Value**
A	368.986 MPa	1097.962 MPa	792 MPa
B	683.973 MPa	1091.964 MPa	510MPa
N	0.73	0.93	0.926
M	1.7	1.1	0.014
C	0.0083	0.014	1.03

Table 3. Failure strain parameters for Al 2024-T3, Ti-6A1-4V and Steel 4340 [17, 22]

Material	**Al 2024-T3**	**Ti-6A1-4V**	**Steel 4340**
Parameter	**Value**	**Value**	**Value**
d_1	0.112	-0.090	0.05
d_2	0.123	0.270	3.44
d_3	1.500	0.480	-2.12
d_4	0.007	0.014	0.002
d_5	0	3.870	0.61

Table 4. Mass of specimen and projectile

Specimen	**Mass (kg)**
Al 2024-T3	0.0283
Ti-6A1-4V	0.0451
Steel 4340	0.0798
Projectile	0.00675

3. Results and discussion

This section discusses the results obtained from the validation of impact damage on the aluminium plate along with the energy absorbed by the plate from both experimentally and numerically. Finally, a parametric study for the nine different configurations mentioned above using three different materials (aluminium, titanium and steel) were carried out and discussed.

3.1. Validation

The results obtained from the experimental ballistic test as well as the numerical analysis are shown in Table 5. It can be observed that the results obtained via FEA agree well with those obtained from the experiments. The small percentage differences observed in the results, i.e. 3.54% and 6.73% for the residual velocity (Vout) and the energy absorbed, respectively, proves that the numerical simulation scheme (FEA) was employed with high accuracy and validity. Figure 4 shows the damage plates after impact for three different initial velocities or gauge pressures.

Table 5. Comparison between numerical and experimental results

Gage pressure (bar)	**V_{in} (m/s)**		**V_{out} (m/s)**		**Energy absorption (J)**	
	Exp.	**FEA**	**Exp.**	**FEA**	**Exp.**	**FEA**
10	155.30	155.30	No penetration	No penetration	-	-
15	191.04	191.04	No penetration	No penetration	-	-
44	272.38	272.38	189.83	183.10	128.77	138.07
			% difference		% difference	
			3.54		6.73	

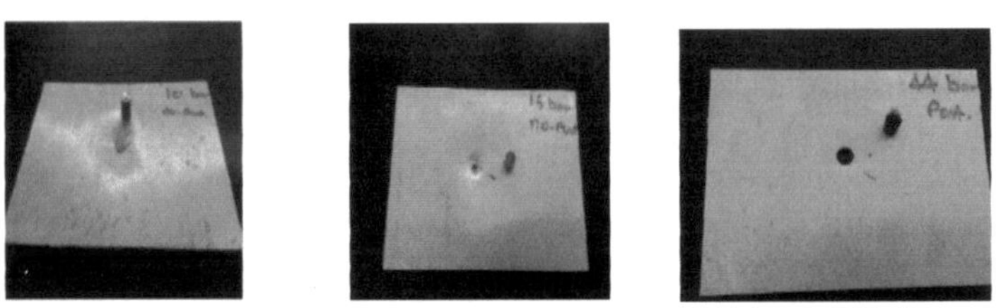

(a) No penetration (b) No penetration (c) Plugging mode

Figure 4. Post-impact damage of Al-2024-T3: (a) V_{in} = 155.3 m/s; (b) V_{in} = 191.04 m/s; (c) V_{in} = 272.38 m/s

3.2. Parametric study

A parametric study using FEA was performed to determine the optimum plate configuration in terms of its energy absorption. From Table 6, it is found that double plate with configuration C7 (Titanium-steel) gives the highest energy absorbed with 279.72 J whereas C5 (Titanium-aluminium) possesses the greatest SEA with 2830.45 J/kg. The high energy absorbed by the C7 plate is attributed the high density of the steel combined with the high strength-to-weight ratio of the titanium alloy. On the other hand, the highest SEA in C5 configuration means impact strength-to-weight ratio of this configuration is the best compared to all other configurations given in Table 1. From this results, it can be concluded that Titanium alloy plays a major role in improving the high velocity impact resistance subjected to blunt projectile especially when used as the front plate facing the projectile. Figures 5 to 8 illustrate the FE results for selected plate configurations under the projectile impact.

Table 6. Results of parametric study using ABAQUS (V_{in} = 500m/s)

Configuration	V_{out} (m/s)	Energy absorbed (Joule)	Specific energy absorbed (energy absorbed/mass of a plate (layer)) J/kg (SEA)
C1	418.9	251.5	1575.714
C2	458.3	134.86	2382.685
C3	423.1	239.54	2655.986
C4	429.0	222.61	1782.225
C5	434.1	207.75	2830.45
C6	442.4	183.205	1694.773
C7	408.8	279.72	2239.551
C8	437.3	198.34	2702.18
C9	412.5	269.47	2492.784

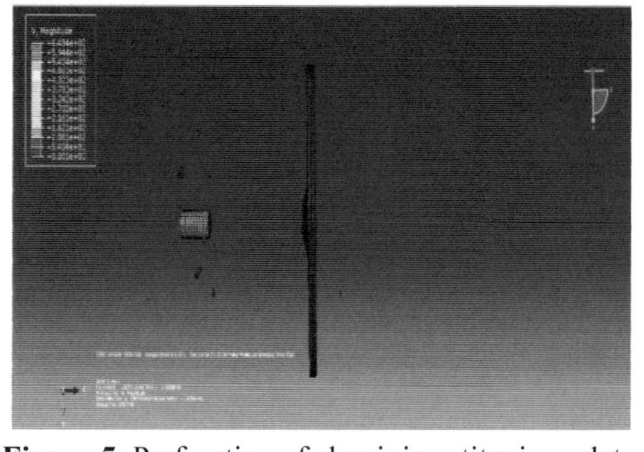

Figure 5. Perforation of aluminium-titanium plate

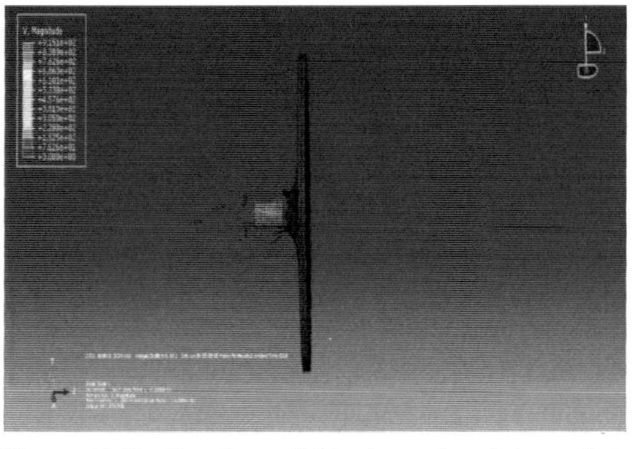

Figure 6. Perforation of titanium-aluminium plate

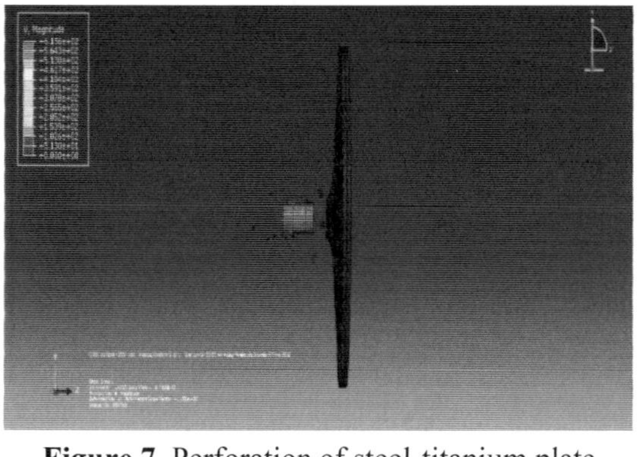

Figure 7. Perforation of steel-titanium plate

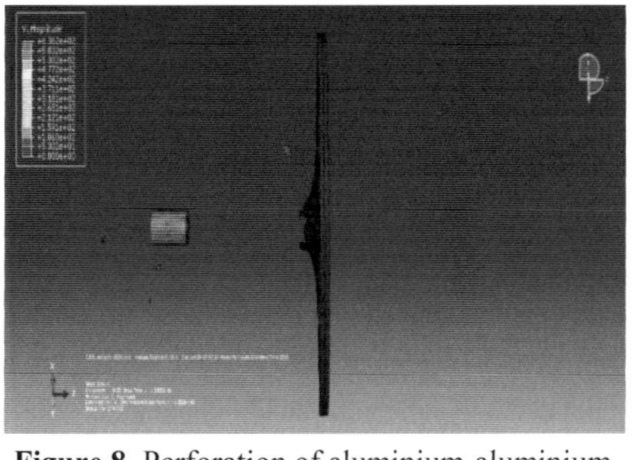

Figure 8. Perforation of aluminium-aluminium plate

4. Conclusion

In this study, the ballistic impact analysis on nine different types of double layered metal plates was investigated by numerical simulation based on the finite element method. Validation of the numerical simulation scheme incorporating the Johnson-Cook damage model was successfully carried out by comparing the numerical results of monolithic aluminium 2024-T3 plate with that of the experimental ballistic test. In the parametric study, it was observed that "titanium-aluminium" specimen yielded the highest specific absorbed energy (SAE) due to the combination of low density (aluminium alloy) and high strength (titanium alloy) characteristics of the double-layered plate. It can be generally concluded that by replacing the either steel or aluminium layer with titanium led to the increase in the SAE.

References

[1] Teng X, Dey S, Borvik T, Weirzbicki T. Protection performance of double-layered metal shields against projectile impact. Journal of Mechanics of Materials and Structures 2007; 2 (7): 1309-1330
[2] Abdelshafy M H, Oyadiji S O. Penetration behaviour of steel plates. ASME International Design Engineering Technical Conferences and 6th International Conference on Multibody Systems, Nonlinear Dynamics, and Control, 2007
[3] Teng T L, Shih T M, Lu C C. Effect of design parameters on the antipenetration properties of space armor. Struct. Eng. Mech 2008; 28(6): 715–725
[4] Zaid A I O, Travis F W. A comparison of single and multi-plate shields subjected to impact by a high speed projectile. Conference of Mechanical Properties at High Rates of Strain, Institute of Physics (UK) Conf. Series, 1971
[5] Nurick G N, Walters C E. The ballistic penetration of multiple thin plates separated by an air gap, SEM Spring Conference on Experimental Mechanics, 1990
[6] Mullin S A, Riegel J P, Tenenbaum D A, Erdley D W. Dynamic plasticity modeling of conical and blunt nosed projectiles and dual layer armor. TACOM Combat Vehicle Survivability Symposium, 1991
[7] Gogolewski R P, Cunningham B J, Riddle R O, Lesuer D R, Syn C K. On the importance of target material interfaces during low speed impact. 16th Int. Symp. on Ballistics, 1996
[8] Jena P K, Ramanjeneyulu K, Kumar K S, Bhat T B. Ballistic studies on layered structures. Mater. Des. 2009; 30:1922–1929
[9] Børvik T, Dey S, Clausen A H. Perforation resistance of five different highstrength steel plates subjected to small-arms projectiles. Int. J. Impact Eng 2009; 36(7): 948–964
[10] Ben-Dor G, Dubinsky A, Elperin T. Applied High-Speed Plate Penetration Dynamics. Dordrecht: Springer, 2006
[11] Ben-Dor G, Dubinsky A, Elperin T. High-Speed Penetration Dynamics. Engineering Models and Methods World Scientific, 2013
[12] Chen H C, Chen Y L, Shen B C. Ballistic resistance analysis of double-layered composite material structures. Theoret. Appl. Fract. Mech. 2012; 62: 15–25
[13] Chen Y L, Chen H C. The numerical method as applied to impact resistance analysis of ogival nose projectiles on 6061–T651 aluminum plates. J. Mech. 212; 28:715–726
[14] Flis L, Sperski M. An investigation of the resistance of multi-layered ships steel shields to 12.7 mm projectiles. Sci. J. Pol. Naval Acad. 2013; 4(195): 31–49
[15] Iqbal M A, Gupta P K, Deore V S, Tak S K, Tiwari G, Gupta N K. Effect of target span and configuration on the ballistic limit. Int. J. Impact Eng 2012; 42: 11– 24
[16] Jankowiak T, Rusinek A, Wood P. A numerical analysis of the dynamic behaviour of sheet steel perforated by a conical projectile under ballistic conditions. Finite Elem. Anal. Des. 2013; 65: 39–49
[17] Failure Modeling of Titanium 6Al-4V and Aluminum 2024-T3 with the Johnson-Cook Material Model. Office of Aviation Research Washington, D.C. 20591, 2003
[18] Saeed M S. Ballistic impact analysis of metallic plates using finite element method. Department

IOP Conf. Series: Materials Science and Engineering **405** (2018) 012012 doi:10.1088/1757-899X/405/1/012012

of Aerospace Engineering, University Putra Malaysia, 2016
[19] US Department of Commerce, National Bureau of Standards, Law Enforcement Standards Laboratory & United States of America. Ballistic Resistant Protective Materials - NIJ Standard 0108.01, 1985
[20] Hibbitt, Karlsson & Sorensen. ABAQUS 6.13 user's manual, 2013.
[21] Johnson G R, Cook W H. A constitutive model and data for metals subjected to large strains, high strain rates and high temperatures. 7th intsymp on ballistics, 1983
[22] Johnson G R, Cook W H. Fracture characteristics of three metals subjected to various strains, strain rates, temperatures and pressures. Engineering Fracture Mechanics 1985; 21: 31-48
[23] Gupta N K, Iqbal M A, Sekhon G S. Experimental and numerical studies on the behavior of thin aluminum plates subjected to impact by blunt- and hemispherical-nosed projectiles. Int J Impact Eng 2006; 32: 1921–1944
[24] Teng X, Wierzbicki T. Evaluation of six fracture models in high velocity perforation. Eng Fract Mech 2006; 73:1653–1678

IOP Conf. Series: Materials Science and Engineering **405** (2018) 012013 doi:10.1088/1757-899X/405/1/012013

Water absorption analysis on impregnated kenaf with nanosilica for epoxy/kenaf composite

F. Bajuri[1], N. Mazlan[1,2*] and M. R. Ishak[1,2]

[1]Department of Aerospace Engineering, Faculty of Engineering, Universiti Putra Malaysia, 43400 UPM Serdang, Selangor.
[2]Institute of Tropical Forestry and Forest Products, Universiti Putra Malaysia, 43400 UPM Serdang, Selangor.

*norkhairunnisa@upm.edu.my

Abstract. Due to the existence of hydroxyl group from cellulose and hemicellulose in natural fibre, it has a high tendency to absorb significant amount of water. Nanosilica has high surface area due to its nanosize, which is capable to seal the fibre's pores. In this work, impregnation technique was used to fill the pores of kenaf fibre with nanosilica in order to reduce the water uptake of the resulting composite. To impregnate the randomly orientated kenaf fibres with nanosilica, the fibres were initially degassed in a plastic container to enable the inclusion of nanosilica to fill the fibre pores. Then nanosilica/epoxy mixture was infused into the container and left under vacuum condition for 24 hours. The hardener was then applied onto the fibres before being hot pressed to aid the curing process. For the water absorption test, the specimens were dried before being placed in a glass container filled distilled water and regularly weighed until approximately 80 days. Like any other kenaf reinforced epoxy, the specimens absorbed water at a rapid rate in the beginning before slowing down and approaching equilibrium. The maximum water absorption rates were 14.4% and 12.9% for specimens with composition of 50 vol% kenaf, 5 vol% nanosilica and 60 vol% kenaf, 5 vol% nanosilica, respectively. Specimens with nanosilica infused but not impregnated had shown a significantly higher water absorption rates at 24.5% and 29.6% respectively. These results show that nanosilica can help reduce the water absorption rate of the natural fibres by blocking the pores of the fibres rendering them unable to absorb water.

1. Introduction

Kenaf, like many other natural plant fibres, has been studied extensively as a susceptible replacement of synthetic fibres such as glass and carbon fibres. Natural fibre is biodegradable, tough, renewable, low cost and low in density, which contribute to its high specific strength and non-abrasive nature, and making natural fibre easy to be processed while allowing high filling levels [1-3]. However, natural fibres tend to absorb moisture either from the air or from rain water, which will affect the properties of the overall composite. The hygroscopic nature of natural plant fibre is contributed by its microscopic like tubes with cell walls surrounding the central lumen [4]. The lumen contributes to water absorption behaviour due to the transport of water molecules via capillary action, which also involves the pores of the fibres [5]. Furthermore, natural plant fibres structure composition contain high amount of cellulose and hemicellulose, which produce the polar hydroxyl group that makes them hydrophilic in nature [6],

[7]. As matrix materials have a tendency of being hydrophobic, hydrophilic natural fibres will repel them and making it difficult to achieve good fibre wettability during production [8].

There are three types of mechanism in which moisture will be absorbed into the composite [9-11] as listed below:

a) Diffusion of the water molecules into the micro bubbles between polymeric chains. The micro bubbles are produced during the mixing and fabrication procedure.
b) Via capillary transport into the gaps and flaws in between the interfaces of fibre and matrix which occurs due to improper fibre wettability and poor fibre-matrix interaction.
c) Micro cracks in the matrix produced due to the swelling of fibres. This mostly happens in the case of natural fibre reinforced composites.

In the case of this work, in addition to the kenaf fibre, nanosilica is introduced into the system and the mixing technique needs to be considered. Basri *et al.* [9] reported that a high speed, low time mixing of nanosilica into epoxy resin system produced specimens with the lowest water absorption rate. When fabricating composite, the idea behind impregnation process is to fill the empty spaces of the natural fibre (lumen cavity and cell wall) with polymer resin [7, 12]. By blocking the water path into the fibre, it is anticipated that the fibre will reduce its water uptake. In this work, nanosilica is considered as the impregnation agent alongside polymer matrix due to its minuscule size contributing to the material's high interface area of about 1500 m^2/g, having active function and high interfacial interaction with resin [9,13]. Nanosilica is expected to be able to impregnate the fibres' lumen to reduce the fibre water absorption rate.

2. Methodology

2.1. Materials

Epoxamite 100 with 103 SLOW Hardener that was manufactured by Smooth-On is used as the matrix. Hydrophilic type of nanosilica derived from rice husk ash was provided by Maero Tech Sdn. Bhd. (MTSB). The reinforcing fibre used was randomly orientated kenaf mat provided by ZKK Sdn Bhd. The kenaf mat was cut into sections of 150 mm × 150 mm and weighed to be utilized in the fabrication process. For every specimen, weighted kenaf mats were combined to achieve the desired weight. It should be noted that nanosilica is an addition to the kenaf and epoxy's loading. Table 1 lists all of the specimens fabricated. In the specimen name, "NI" after the number indicates that the specimen is not impregnated whereas "I" is designated for impregnated specimens.

Table 1. List of specimens fabricated

Specimen Name	Kenaf Loading [vol%]	Epoxy Loading [vol%]	Nanosilica Loading [vol%]	Impregnated or not
400NI	40	60	0	No
500NI	50	50	0	No
600NI	60	40	0	No
405NI	40	60	5	No
505NI	50	50	5	No
605NI	60	40	5	No
405I	40	60	5	Yes
505I	50	50	5	Yes

IOP Conf. Series: Materials Science and Engineering **405** (2018) 012013 doi:10.1088/1757-899X/405/1/012013

2.2. Fabrication of non-impregnated specimens

The nanosilica is homogenized with epoxy without the hardener at 3000 rpm for 10 minutes using a homogenizer. For the specimens without nanosilica, this homogenization procedure was skipped. The hardener was then mixed into the solution and applied evenly using a brush onto the cut kenaf mats. The wet kenaf mats were stacked inside a 150 mm × 150 mm × 3 mm mould. The mould was then compressed using a 40 tonne hot press at pre-heated at 85°C for 20 minutes and then subsequently cold pressed for 5 minutes. The specimens were left overnight for curing. All specimens were post cured at 80°C for 2 hours.

2.3. Fabrication of impregnated specimens

The nanosilica is homogenized into approximately 2.5 litres of epoxy without the hardener using a homogenizer at 5000 rpm for 20 minutes. The speed and time of the homogenizing process were faster and longer than non-impregnated specimens due to the amount of matrix and nanosilica used. Inside a bell jar, a container with kenaf mats was placed. The air inside the glass jar was evacuated to an impregnation pressure of 600 mmHg (80 kPa) for 5 minutes using a hydraulic vacuum pump before the epoxy and nanosilica were infused until the kenaf mats were fully submerged. The specimens were left under impregnation pressure of 600 mmHg for another 2 hours before the pump was switched off. The specimens were left overnight before releasing the air into the vessel slowly within 30 seconds. The impregnated kenaf mats were drained off by compressing the fibres. After that, the hardener was applied using a brush and the kenaf mats were placed inside a 150 mm × 150 mm × 3 mm waxed mould before being hot pressed at 85°C for 5 minutes and subsequently cold pressed for 5 minutes. The specimens were left overnight for curing. All specimens were post cured at 80°C for 2 hours.

2.4. Water absorption test

The water absorption test was conducted according to ASTM D570-98. The samples were conditioned into approximately 57 mm × 10 mm. For each loading, 3 replications were made. The samples were dried in an oven at 40°C for 24 hours before being weighed. Then the samples were immersed into distilled water with controlled temperature of 30 ± 1°C. Samples were then weighed subsequently after 3, 6, 24, 48, 72 and 96 hours for the first week. Further weighing was conducted every one to two weeks for approximately two and a half months. The weighing is done by removing the sample from water and wiping excess water with dry cotton cloth before placing the sample onto a measuring device with the accuracy to the nearest 0.001 g. The sample was then replaced back into the water immediately. The distilled water was changed every one to two weeks.

Water intake of each sample was calculated using percentage weight gain of the specimens at the interval measured as shown by Eqn. 1, where M_i is the percentage weight gain, M_d is the initial oven dried mass and M_t is the mass at the interval measured.

$$M_i = \frac{M_t - M_d}{M_d} \times 100 \qquad (1)$$

3. Results and discussion

Figure 1 shows water absorption behaviour of nanosilica/kenaf reinforced epoxy (SKRE) and kenaf reinforced epoxy (KRE). For the first day, the KRE absorbed water rapidly at an almost linear rate. This was true for all specimens. Then, the water absorption rate (weight percentage increase rate) slowed down gradually with time. This could be seen until day 9 (15 root hour) for the impregnated specimens and day 25 (24 root hour) for the not impregnated specimens. Further immersions afterward only showed a slight weight increase as the specimens were nearing the equilibrium. Water molecules penetrated the cell walls and occupy the space in fibre lumen, causing the fibre density to increase [7]. In other words, the fibre will swell due to moisture intake. Fibre swelling results in the production of voids in between fibre and matrix interface, furthering penetrability of water and inducing capillary action [14]. Therefore, it can be concluded that rapid water uptake into the composite in the beginning is governed by multiple mechanisms: water absorbed by the fibres, voids induced by fibre swelling and existence of micro porosity in the matrix system. When the weight of the specimens are almost constant, it shows that the specimens have less interaction sites with water molecules and is getting saturated with water [9]. All specimens show similar absorption trend like that of Fickian diffusion trend where the water is absorbed sharply and then the absorption rate decreased until the specimens' weight percentage increase reaches equilibrium.

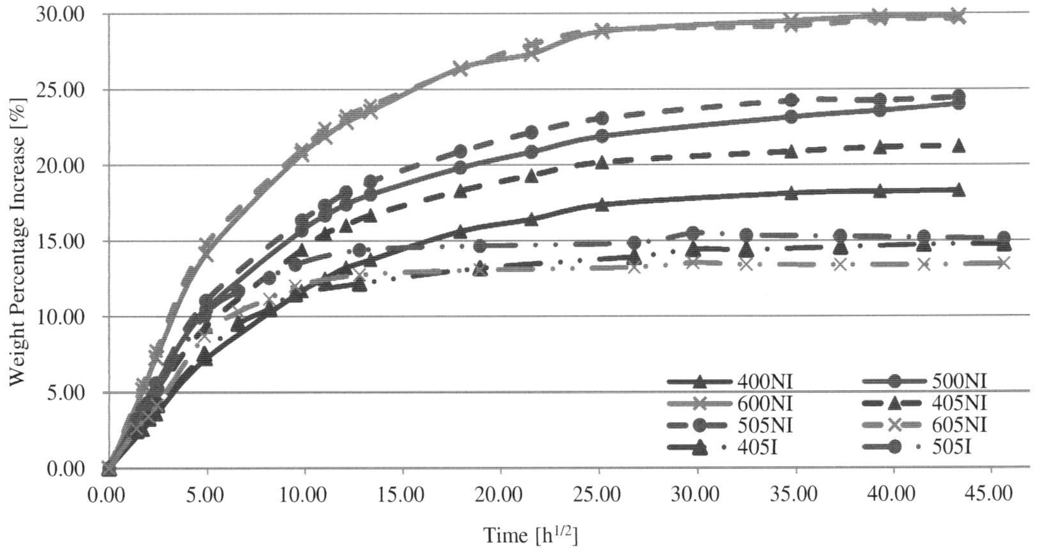

Figure 1. Water absorption curves of SKRE and KRE

Table 2 shows the maximum water weight percentage increase for every specimen. From the table, it can be observed that with the increment of kenaf, the maximum water intake is increased too. This finding is in tandem with the knowledge where, the higher content of hydrophilic fibre in a composite, the higher the amount of water will be absorbed. Furthermore, with the exception of 605NI specimen, inclusion of 5 vol% nanosilica has slightly increased the water absorption rate. During homogenizing process, air bubbles may be produced too. The air bubbles consequently are excess spaces for water to occupy inside the composite. In another words, inclusion of nanosilica produces more micro bubbles where water will be diffused into, hence increasing the water absorption rate of a composite [9-11]. Furthermore, improper homogenizing may cause nanosilica to agglomerate, thereafter producing even more voids as the spaces in between nanoparticles are occupied with air instead of the matrix.

Table 2. Maximum weight percentage increase of SKRE and KRE

Specimen	Maximum Weight Increase [%]
400NI	18.29
500NI	24.00
600NI	29.80
405NI	21.21
505NI	24.45
605NI	29.63
405I	14.74
505I	15.10
605I	13.44

However, with the impregnation conditioning, the specimens' maximum weight increase after the immersion in distilled water was considerably lower than the not impregnated specimens, especially for 605I specimen. There are multiple reasons on the impregnated specimen low water absorption rate such as listed as follow:

a) The absence of micro bubbles. Previously, it is mentioned that homogenizing of nanosilica into epoxy produces more air bubbles. During the impregnation process, the matrix and fibres are subjected to vacuum pressure overnight. Apart from filling the lumen of fibres with nanosilica, this process helps in removing micro bubbles too.
b) The impregnation process succeeds in filling the lumen cavity and cell wall with epoxy and nanosilica. The matrix and nanoparticles that impregnated the fibres' cell wall and cell lumen acted as a wall that blocked water molecules from diffusing into the fibres or at least reduced the permeability of behaviour of the fibre as the micro pores were blocked due to the matrix existence [7].
c) Good interfacial adhesion between kenaf fibres and epoxy. Poor adhesion between matrix and the lignocellulosic fibres will cause more gaps to appear in between the interfacial region making more hydrogen bonding occur between waterand higher abundance of hydroxyl group from the fibres [15]. Epoxy resin used in this work takes 1 hour before the gelling process. This means that when applying epoxy, there is only approximately 1 hour for the matrix to properly wet the fibres. After 1 hour, due to the gelling process, it will be harder for the matrix to create adhesion with the fibres due to the increased viscosity. With impregnation, as the fibres are submerged for overnight inside the resin, there are more time for the matrix to create interfacial adhesion with the fibres.

4. Conclusion

Higher fibre volume content will increase the KRE absorption of more moistures and the inclusion of nanosilica too will increase the water absorption rate. Impregnation process considerably reduces the water absorption rate. 605I specimen has the lowest water uptake at 13.44% weight increase from the initial submersion. The low water uptake indicates that the nanosilica/epoxy mixture has succeeded in penetrating the cell walls and lumen of the fibre, subsequently blocking water from being absorbed.

Acknowledgement

The authors would like to thank Universiti Putra Malaysia (UPM) for providing the grant to further this research. Furthermore, the authors would like to thank FRIM (Forest Research Institute Malaysia) for providing facility to conduct impregnation experiment. The authors also like to acknowledge Universiti Putra Malaysia for providing the Putra research grant 9405401, 9405400 and 9463900.

References

[1] Marques M, Melo R, Araujo R, Lunz J, Aguiar V. Improvement of mechanical properties of natural fiber-polypropylene composites using successive alkaline treatments. J. Appl. Polym. Sci. 2015; 132(12): 1–12
[2] Kim S J, Moon J B, Kim G H, Ha C S. Mechanical properties of polypropylene/natural fiber composites: Comparison of wood fiber and cotton fiber. Polym. Test. 2008; 27(7): 801–806
[3] Lee B H, Kim H S, Lee S, Kim H J, Dorgan J R. Bio-composites of kenaf fibers in polylactide: Role of improved interfacial adhesion in the carding process. Compos. Sci. Technol. 2009; 69(15–16): 2573–2579
[4] Thomas S, Paul S, Pothan L A, Deepa B. Natural fibres: Structure, properties and applications. in Cellulose Fibers: Bio- and Nano- Polymer Composites, Berlin Heidelberg: Springer-Verlag, 2011
[5] Megiatto J D, Silva C G, Ramires E C, Frollini E. Thermoset matrix reinforced with sisal fibers: Effect of the cure cycle on the properties of the biobased composite. Polym. Test. 2009; 28(8): 793–800
[6] Akil H M, Omar M F, Mazuki A, Safiee S, Ishak Z A M, Abu Bakar A. Kenaf fiber reinforced

composites: A review. Mater. Des. 2011; 32(8–9): 4107–4121

[7] Ishak M R, Sapuan S M, Leman Z, Rahman M Z A, Anwar U M K, Siregar J P. Sugar palm (Arengapinnata): Its fibres, polymers and composites. Carbohydr. Polym. 2013; 91(2): 699–710

[8] Rassmann S, Reid R G, Paskaramoorthy R. Effects of processing conditions on the mechanical and water absorption properties of resin transfer moulded kenaf fibre reinforced polyester composite laminates. Compos. Part A Appl. Sci. Manuf. 2010; 41(11): 1612–1619

[9] Basri M S M, Mazlan N, Mustapha F. Effects of stirring speed and time on water absorption performance of silica aerogel/epoxy nanocomposite. ARPN Journal of Engineering and Applied Sciences 2015; 10(21): 9982–9991

[10] Akil H M, Cheng L W, Mohd Ishak Z A, Abu Bakar A, Abd Rahman M A. Water absorption study on pultruded jute fibre reinforced unsaturated polyester composites. Compos. Sci. Technol. 2009; 69(11–12): 1942–1948

[11] Dhakal H N, Zhang Z Y, Richardson M. Effect of water absorption on the mechanical properties of hemp fibre reinforced unsaturated polyester composites. Compos. Sci. Technol. 2007; 67(7–8): 1674–1683

[12] Ishak M R, Leman Z, Salit M S, Rahman M Z A, Anwar Uyup M K, Akhtar R. IFSS, TG, FT-IR spectra of impregnated sugar palm (Arengapinnata) fibres and mechanical properties of their composites. J. Therm. Anal. Calorim. 2013; 111(2): 1375–1383

[13] Bajuri F, Mazlan N, Ishak M R, Imatomi J. Flexural and compressive properties of hybrid kenaf / nanosilica in epoxy composite. Procedia Chem. 2016; 19: 955–960

[14] Mazuki A A M, Akil H M, Safiee S, Ishak Z, Bakar A A. Degradation of dynamic mechanical properties of pultruded kenaffiber reinforced composites after immersion in various solutions. Compos. Part B Eng. 2011; 42(1): 71–76

[15] Nourbakhsh A, Baghlani F F, Ashori A. Nano-SiO_2 filled rice husk/polypropylene composites: Physico-mechanical properties. Ind. Crops Prod. 2011; 33(1): 183–187

Thermogravimetric analysis on rice husk ashes-based geopolymer paste

N R A Nasruddin[1], N Mazlan[1,2*], M S Mohd Basri[3], A H Mohamed Ariff[2,4] and N A Mohd Nasir[5]

[1]Department of Aerospace Engineering, Faculty of Engineering, Universiti Putra Malaysia, 43400 UPM Serdang, Selangor.
[2]Institute of Tropical Forestry and Forest Products, Universiti Putra Malaysia, 43400 UPM Serdang, Selangor.
[3]Department of Process and Food Engineering, Faculty of Engineering, Universiti Putra Malaysia, 43400 UPM Serdang, Selangor.
[4]Department of Mechanical and Manufacturing Engineering, Faculty of Engineering, Universiti Putra Malaysia, 43400 UPM Serdang, Selangor.
[5]Department of Civil Engineering, Faculty of Engineering, Universiti Putra Malaysia, 43400 UPM Serdang, Selangor.

*norkhairunnisa@upm.edu.my

Abstract. Geopolymer is a material produced from the reaction of rice husk ashes and alkaline solution. These rice husk ashes are not widely used yet due to unstable chemical content than fly ashes. However, the high Si content in rice husk ashes are believed to contribute for better thermal properties such as in coating application. This paper will discuss the studies done to rice husk ashes geopolymer in aspects of thermal stability via conducting thermogravimetric analysis. Thermogravimetric analysis is carried out to determine the optimum temperature for each sample. This work shows the reactions of geopolymer paste with different molarity of NaOH solutions.

1. Introduction

Geopolymer is a binding material produced from reaction between rice husk ash (RHA) and alkaline activator (AA) solution. RHA is known as a cheap raw product material with good material properties that can be utilized for many applications [1]. Apart from that, RHA is also known as one of the most available wasted biomass resources. Due to its low cost price and biodegradable characteristic, RHA has become one of the most potential materials to be applied for engineering application. Combination of different alkaline activators such as sodium silicate (Na_2SiO_3) and sodium hydroxide (NaOH) can create synergistic effects and highly potential in dissolving the rice husk ash to produce a geopolymer binder [2-3]. Geopolymers have been known to have high mechanical properties, low shrinkage, low permeability and good thermal properties [2-4]. To investigate a process that involves many potential factors to obtain optimum result is tiresome and time consuming, hence works from other researchers are referred to. Many have agreed that the three most influencing factors in geopolymer materials are ratio solid to liquid, ratio Na2SiO3 to NaOH and NaOH molarity [5-8].

IOP Conf. Series: Materials Science and Engineering **405** (2018) 012014 doi:10.1088/1757-899X/405/1/012014

The objective of this study is to analyze the important factors that mostly control the weight loss of RHA based geopolymer material. Furthermore, to assist in clarifying the effects of each factor on the material properties, a testing called thermogravimetric analysis is done.

2. Experimental work

2.1. Preparation of material

The material used is rice husk ash (RHA), which is grounded to a finer size of particles. Then RHA particles is sieved to the average size of 50 microns. Table 1 shows the sample designation based on the composition made for each sample. The chemical composition of the rice husk ash is tabulated in Table 2, which is obtained from the X-ray Flourescence (XRF) analysis. The chemical composition for sodium silicate (Na_2SiO_3) is comprised of 57.8% SiO_2, 11.9% Na_2O and also 30.3% H_2O. Sodium hydroxide (NaOH) pellets with 98% purity are used after being diluted using distilled water to obtain 4M, 8M and 12M.

Table 1. List of samples

Sample	Ratio RHA to AA	Ratio Na_2SiO_3 to NaOH	NaOH Molarity
S01	1.0	0.5	4
S02	1.5	0.5	4
S03	1.0	2.5	8
S04	1.5	0.5	8
S05	1.0	0.5	12
S06	1.5	0.5	12

Table 2. Chemical composition of rice husk ash

Compound	Concentration Unit (%)
Al_2O_3	3.00
SiO_2	87.40
K_2O	0.49
CaO	1.40
Cr_2O_3	0.27
MnO	0.19
Fe_2O_3	1.49
NiO	0.07

2.2. Mixing process

Initially, the alkaline activator (AA) solution is prepared by combination between NaOH and Na_2SiO_3. The preparation of geopolymer solution is done by mixing the RHA particles with the designated ratio of alkaline activator solution. The mixtures are then mechanically stirred for 5 to 10 minutes until they are homogeneous. The homogeneous mixture of geopolymer solution are then cast onto moulds with the size of 70mm x 50mm x 12mm. Finally, samples are left cured at 75°C for 24 hours in the oven.

2.3. Thermogravimetric analysis

The thermogravimetric analysis was done using Pyris 1 TGA instrument and samples were prepared according to the required standard. The samples were prepared to be weighed less than 5 mg cube. The instrument was set to heat from 30°C to 500°C at 20.00°C/min.

3. Results and discussion

This paper analyses the thermal behaviour on geopolymer samples and the influencing factors that are affecting the weight loss of the geopolymer samples. A well-combined sodium silicate and sodium hydroxide were used as the alkaline activator (AA). The reason of using these two solutions because they are cheaply available and conveniently to be used [6, 9-10]. There are a total of six samples that were thermally tested as shown in Table 3.

Table 3. Percentage of weight loss and mass ratio for all samples

Sample	**Percentage of Weight Loss (%)**		**Molar Ratio**			
	150°C	300°C	Na_2O/SiO_2	SiO_2/Al_2O_3	H_2O/Na_2O	water/ solid
S01	29	49	0.105	60.169	20.736	1.933
S02	10	15	0.074	56.791	20.719	1.408
S03	32	64	0.108	72.593	10.669	1.027
S04	19	49	0.044	56.791	30.543	1.251
S05	25	54	0.206	60.169	8.853	1.489
S06	18	38	0.146	56.791	8.853	1.109

In the TGA test, the weight loss was measured while the geopolymer samples were gradually exposed to the increasing temperature. Figure 1 represents the results of the TGA analysis on the percentage of weight loss of the geopolymer samples. It shows in overall that the decreased in weight starts before 100°C as recorded in the TGA analysis, which can be associated with the loss of evaporable water in the geopolymer as claimed by Al Bakri *et al.* [7]. Furthermore, all the geopolymer samples stabilized the weight loss between 200°C to 500°C.

Meanwhile Figure 2 shows the pattern of degradation of the RHA based geopolymer samples are similar, which means that the samples start to degrade before they reach 100°C. The decomposition of the RHA based geopolymer samples took place during 100°C to 250°C. The decomposition process of the samples stop as the temperature when higher than 250°C.

3.1. Effect of ratio solid to liquid and NaOH molarity

According to the Figure 1(a), which is 4M for both samples, at 150°C, S01 losses 29% and S02 losses only 10%. As the temperature goes to 300°C, S01 sample losses its weight for about 49% while for S02 sample, the weight loss is about 15%, which is increased by 5% only. Furthermore, in Figure 1(b) for 8M samples, at 150°C: S03 losses 32% and S04 losses 19% and at 300°C: S03 losses 64% and S04 losses 49%. However, the increment trend for both samples to lose weight as the temperature went from 150°C to 300°C is similar. Lastly, Figure 1(c) is for 12M samples, at 150°C: S05 losses 25% and S06 losses 18% and at 300°C: S05 losses 54% and S06 losses 38%. For 12M samples, S05 has higher losses than S06 in weight as the temperature went higher.

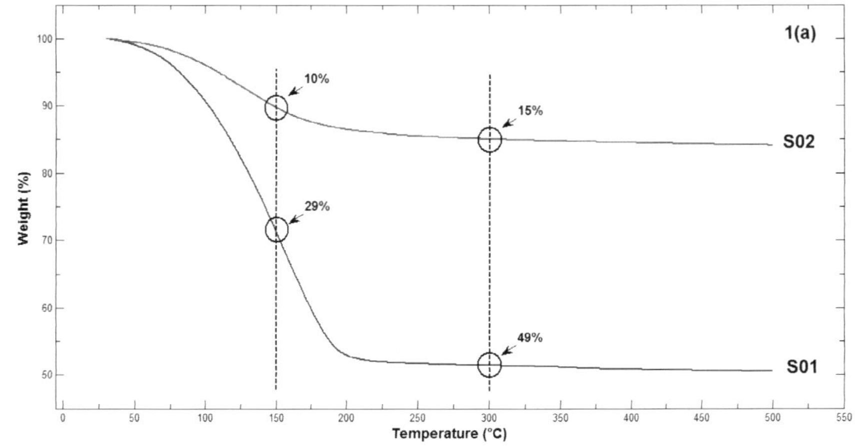

(a)

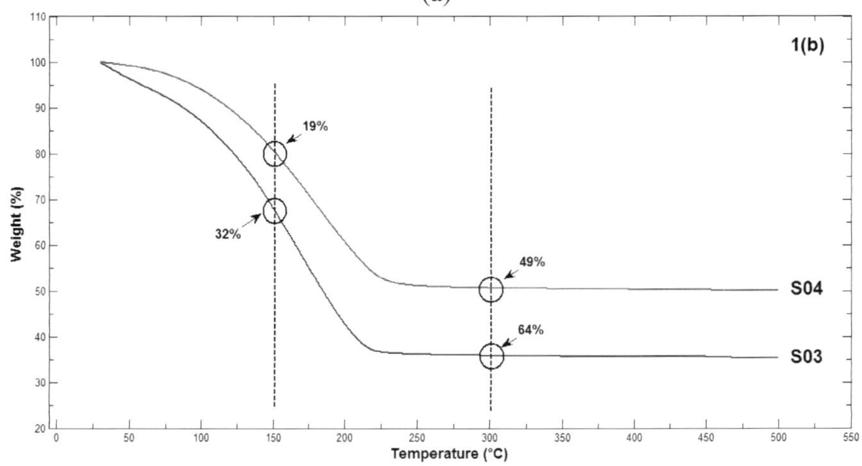

(b)

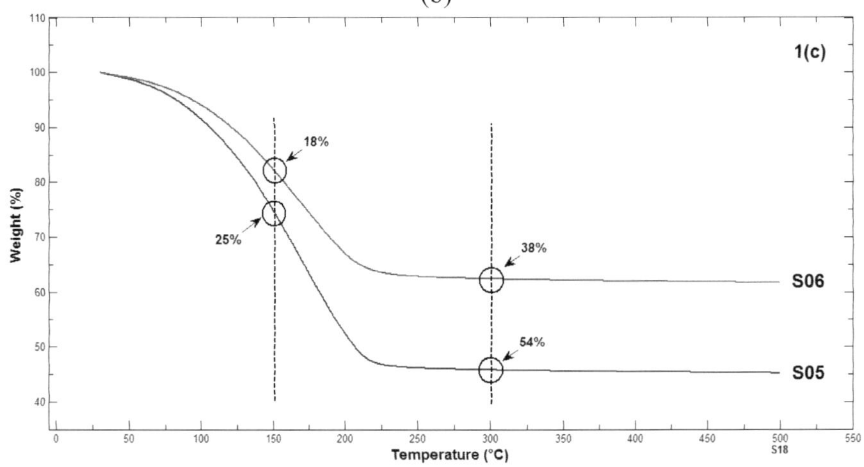

(c)

Figure 1. TGA graphs for every molarity: (a) 4M of NaOH molarity, (b) 8M of NaOH molarity, (c) 12M of NaOH molarity

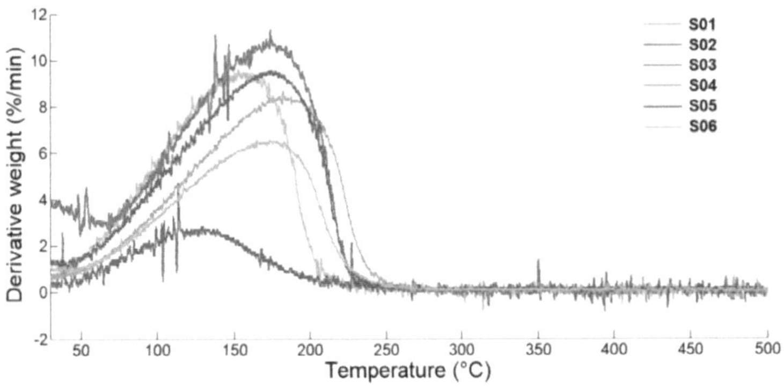

Figure 2. DTGA graphs for every molarities of RHA based geopolymer

The calculated ratio of solid to liquid is based on mass ratios (kg RHA per kg AA) as well as the ratio of Na_2SiO_3 to NaOH (kg Na_2SiO_3 per kg NaOH) for all samples. Ratio 1.0 of solid to liquid contents about 50% of RHA whereas in ratio 1.5 of solid to liquid contents 60% of RHA. That means these 1.5 ratio contents have more RHA than 1.0 ratio, thus the bonding structure of geopolymer more compact and less porous. S01, S03 and S01 are 1.0 ratio of solid to liquid whereas S02, S04 and S06 are 1.5 ratio of solid to liquid. During fabrication, geopolymer samples with the ratio 1.0 have better workability and flow ability during the mixing process because of its low RHA content than ratio 1.5. Geopolymer with low RHA content is much easier to stir since it has more water content than solid. Since the samples with ratio 1.0 solid to liquid have the highest water content, they probably require longer curing time or increase of curing temperature to release water binders and to reduce adsorbing more moisture [11-13].

3.2. Effect of ratio water/solid and Si/Al

The molar ratio is a ratio of main compounds available in every sample. Factors such as high ratio of Si/Al and low water content samples were found to produce better thermal properties [1-2, 11]. S03 has highest ratio of Si/Al and lowest ratio water/solid. But according to the weight loss graph, S03 has the most highest in weight loss about 32% from 150°C to 300°C than other samples. For instance, S02 and S04 have higher ratio water/solid than S03 and they have similar value of ratio Si/Al that is less than S03. However, some researchers have claimed that the ratio of water/solid, ranging between 1.35 and 1.50, shows better results but it has low workability and caused inhomogeneous geopolymer [1].

4. Conclusion

This paper discusses the thermal behaviour for the RHA geopolymer samples, which is also influenced by the Si/Al content. Sample S03 has the lowest ratio water/solid and highest ratio SiO_2/Al_2O_3, thus is potentially a better thermal stability RHA geopolymer. Despite that, S03 has the highest weight loss during thermal analysis compared to S02 that is able to maintain 75% of its weight. Moreover, the ratio water/solid for S02 is still in the range of 1.30 and 1.50 that researchers have claimed to be the better workability geopolymer [1, 15]. Still, all geopolymer samples in this study show good thermal stability because the amount of Si/Al in every sample is high.

Acknowledgment

The authors would like to acknowledge the financial support of Fundamental Research Grant Scheme (FRGS) from Department of Higher Education, Malaysia (Vot. Number: 5524464) and also Universiti Putra Malaysia (UPM) for funding this research study under Putra Grant (Vot. Number: 9405401).

References

[1] Mohd Basri M S, Mustapha F, Mazlan N, Ishak M R. Fire Retardant performance of rice husk ash-based geopolymer coated mild steel - A factorial design and microstructure analysis.

Mater. Sci. Forum 2016; 841: 48–54
[2] Cheng Yong H, Yun Ming L, Al Bakri Abdullah M M, Hussin K. Fire resistant properties of geopolymers: A review. Key Eng. Mater. 2015; 660: 39–43
[3] Salwa M S S, Al Bakri A M M, Kamarudin H, Ruzaidi C M, Binhussain M, Sharifah Zaliha S Z. Review on current geopolymer as a coating material. Australian Journal of Basic and Applied Sciences 2013; 7(5): 246–257
[4] Ramujee K, Potharaju M. Abrasion resistance of geopolymer composites. Procedia Mater. Sci. 2014; 6: 1961–1966
[5] Farhana Z F, Kamarudin H, Rahmat A, Al Bakri A M M. The relationship between water absorption and porosity for geopolymer paste. Mater. Sci. Forum 2014; 803: 166–172
[6] Hardjito D, Cheak C C, Ho C, Ing L, Ing C H L. Strength and setting times of low calcium fly ash-based geopolymer mortar. Mod. Appl. Sci. 2008; 2(4): 3–11
[7] Mustafa Al Bakri A M, Kamarudin H, Binhussain M, Rafiza A R, Zarina Y. Effect of Na_2SiO_3/ NaOH ratios and NaOH molarities on compressive strength of fly-ash-based geopolymer. ACI Mater. J. 2012; 109(5): 503–508
[8] Hardjito D, Rangan B V. Geopolymer concrete for environmental protection. Res. Rep. GC 2014; 41–59
[9] Mustafa Al Bakri A M, Kamarudin H, Binhussain M, Khairul Nizar I, Zarina Y, Rafiza A R. The effect of curing temperature on physical and chemical properties of geopolymers. Phys. Procedia 2011; 22: 286–291
[10] Al Bakri A M M, Kamarudin H, Binhussain M, Nizar I K, Rafiza A R, Zarina Y. Microstructure of different NaOH molarity of fly ash- based green polymeric cement. J. Eng. Technol. Res. 2011; 3(2): 44–49
[11] Riahi S, Nazari A, Zaarei D, Khalaj G, Bohlooli H, Kaykha M M. Compressive strength of ash-based geopolymers at early ages designed by Taguchi method. Mater. Des. 2012; 37: 443–449
[12] Hardjito D. *Studies on Fly Ash-Based Geopolymer Concrete*. Curtin University of Technology, 2005.
[13] Deb P S, Nath P, Sarker P K. The effects of ground granulated blast-furnace slag blending with fly ash and activator content on the workability and strength properties of geopolymer concrete cured at ambient temperature. Mater. Des. 2014; 62: 32–39
[14] Temuujin J, Minjigmaa A, Rickard W, Lee M, Williams I, van Riessen A. Preparation of metakaolin based geopolymer coatings on metal substrates as thermal barriers. Appl. Clay Sci. 2009; 46(3): 265–270
[15] Temuujin J, Minjigmaa A, Rickard W, Lee M, Williams I, van Riessen A. Fly ash based geopolymer thin coatings on metal substrates and its thermal evaluation. J. Hazard. Mater. 2010; 180(1–3): 748–752

IOP Conf. Series: Materials Science and Engineering **405** (2018) 012015 doi:10.1088/1757-899X/405/1/012015

Robust baseline design selection methodology: aircraft redesign case study

Fairuz I Romli*, Nur Shafiqah Daud and Nadia Hanani Ahmad Habibillah

Department of Aerospace Engineering, Faculty of Engineering,
Universiti Putra Malaysia, 43400 Serdang, Selangor, Malaysia.

fairuz_ir@upm.edu.my

Abstract. In product redesign process, the chosen baseline design plays an important role in determining the success of the development process. A good baseline design has a suitable level of architectural flexibility to absorb the change effect without propagating it beyond the initiating change components. Few methods for baseline design selection have been proposed but none of them consider the uncertainties of the change impact level during the assessment process. This study presents a robust baseline design selection method through a simple aircraft redesign case study that includes the consideration of change impact variation. The resultant distribution of the overall evolvability risk score for the baseline candidate can provide more insights on its suitability for the redesign task at hand. This allows a better selection of a more robust baseline design that lead a higher probability of product development success.

1. Introduction

Most available products in the market today are essentially the improved or upgraded versions of their previous predecessor designs. In other words, many of today's product manufacturers rarely start their design and development process from scratch. Instead, they utilize existing designs as the modification basis for improving them to be close to the new customer demands. For instance, aircraft or jet engines are exemplary complex products that have evolved from generation to generation through the transfer and revision of their design elements, ranging from general solution principles to details of component manufacture [1]. It has been conservatively approximated that more than 75% of design activities can be categorized as design modification, variant design or case-based design [2]. This highlights the high adaptation of redesign process in current product development, which is heavily driven by the market competition. Unlike a totally new product design and development, redesigning successful products by reusing their proven design elements and solution principles enables the process to be done faster and helps to leverage costs and risks [3]. Many redesigned products, although do not deviate far from their predecessor designs, are typically still perceived as novel by the customers.

One of the main differences between totally new and redesign product development is the existence of a well-defined baseline product design for the latter process. The complexity of the redesign process is closely tied to the appropriateness of the selected baseline design to the driving requirements. It has been demonstrated in a previous study that different baseline aircraft designs will have different level of redesign risks for similar development goals because of their different system architectures [4]. This particular condition is more pronounced in complex products since their design architecture is highly interrelated. The redesign process involves making modifications to the baseline design and due to the interrelationships between its parts or components, either physically or functionally, the effects of the changes made can be propagated to other parts or components within the system design architecture [5]. This phenomenon is known as "engineering change propagation" and it will adversely affect the budget and scheduling constraints of the redesign process. A good example on the negative impacts of the change propagation can be observed from the helicopter redesign case in the Westland Helicopters Company. From what should have been a simple redesign change to add the forward looking infrared radar (FLIR) turret, it has caused further modifications to its avionic, fuselage structure and nose cap, power supply, cabling and piping [6]. If the change propagation effects are underestimated, the impact to the product redesign development can be severe [7]. In general, the probability of occurrence of the change effects propagation is highly reliant on the level of connectivity between parts or components of the system design architecture and for redesign process, the interrelationships are very much well-defined for the existing baseline product design. For that reason, to reduce the effects of the change propagation, a better selection of the baseline design has to be made.

2. Previous works on baseline design selection methods

It is believed that baseline designs for redesign process are mostly selected based on their closeness to the target requirements or because they are the natural choice for incremental progression within their product family [8]. Such selection presumes that the imminence of the baseline's current capabilities to the driving requirements will ensure a minimum amount of required changes but, as the propagation of change effects goes, this assertion is not always true. A minor modification on existing product design can cause propagated changes to many other parts of the product while making a few major initiating changes may be limited to only the particular initial parts. This entails a careful look into the existing product architecture design in the selection of a good baseline design for redesign process and not just evaluating on the proximity of its current system performance to the target requirements.

The earliest work of the authors on baseline design selection comes as part of the development for the Strategic Planning of Engineering Changes (SPEC) methodology [9]. Realizing that the amount of changes to be made on the baseline product system also depends on its design architectural flexibility, several assessment metrics have been proposed to measure its goodness as the baseline for the specific redesign purposes. It should be noted that the suitability or goodness of the product system to become the baseline design varies for different redesign goals. The metrics: generality, scalability, adaptability, extensibility and complexity, have been primarily defined in reference to system evolvability measures in Ref. [10] and Methodology for Assessing the Adaptability of Products (MAAP) [11]. The definition of the baseline evaluation metrics is tabulated in Table 1. An example case of baseline assessment as a part of the SPEC methodology framework for aircraft system redesign process has been presented in Ref. [12], where it has been demonstrated with the use of baseline evaluation method that different baseline aircraft candidates have different level of redesign risks.

Table 1. Definition of baseline design assessment metrics [8]

Metric	Description
Generality	Capacity to accommodate changed or new requirements without requiring any changes to its existing design
Scalability	Capacity to accommodate required changes only by the scaling of its existing design without requiring any new components
Adaptability	Capacity to accommodate required changes without propagating the change effects beyond the initiating components
Extensibility	Capacity to accommodate required changes with the effects propagation allowed
Complexity	Capacity to accommodate required changes without increasing its design complexity level

IOP Conf. Series: Materials Science and Engineering **405** (2018) 012015 doi:10.1088/1757-899X/405/1/012015

Using the same evaluation approach and principles, several assisting decision-making methods for product redesign process have been developed. These include Strategic Change Ranking Methodology (SCRaM) [13] and Change Ranking of Product Subsystems (CROPS) [14]. The work presented in this paper is essentially the continuation from a previous study in Ref. [15], where the Monte Carlo method is applied in the prediction of the subsystem redesign risk for complex products. The use of the Monte Carlo method here is to include the consideration of the uncertainty surrounding whether the redesign modifications made will introduce any propagated change effects to the adjacent parts or components. However, there are also uncertainties regarding the level of impact caused by the redesign changes that are not being considered. In previous studies, the change impact level is taken as a static value that has been predicted based on past design experiences, which might not be correct most of the times. Hence in this presented study, the change impact level is considered as a variable that has uncertain value and the resultant baseline product design from such assessment method is believed to be more robust with this additional consideration.

3. Robust baseline design selection

In general, the efficiency of the product redesign process can be estimated based on the amount of cost and efforts that it requires, which is translated into process risks when the redesigned product fails to meet all of the desired requirements. A widely-used scheme in risk management approximates change risk as the product of its likelihood and change rating [16, 17]. With this in mind, the prediction of the change impact from the product redesign works can be made through the translation of their level of difficulty and cost to realize them. Furthermore, the difficulty to implement the redesign changes can be related to the readiness of the technology to realize them. Based on this notion, the rating for level of change difficulty can be based on the definition of System Readiness Level (SRL). SRL is a good reference for qualitative change risk rating because it provides a great reflection of the change process at hand by relating to the system or technology readiness to be incorporated into the product's use [18]. Table 2 and Table 3 present the rating scale used for change difficulty and cost, respectively.

Table 2. Change difficulty rating scale

Rating	Description
1	The component modification is at SRL 5
4	The component modification is at SRL 4
6	The component modification is at SRL 3
8	The component modification is at SRL 2
10	The component modification is at SRL 1

Table 3. Change cost rating scale

Rating	Description
1	Very low cost
4	Low cost
6	Medium cost
8	High cost
10	Very high cost

Based on Ref. [8], the baseline design assessment metrics can be evaluated using the following Eqn. 1 to Eqn. 5. As can be observed from the equations, the inclusion of uncertainties with regards to change impacts will only affect the risk of adaptability and extensibility metrics whereas the risk of generality, scalability and complexity metrics remain as static values depending on the assessment of the redesign task at hand.

$$Generality\ Risk = \sum \begin{cases} 0 & no\ change \\ 100 & require\ change \end{cases} \tag{1}$$

$$Scalability\ Risk = \sum \begin{cases} 0 & no\ new\ component \\ 100 & require\ new\ component \end{cases} \tag{2}$$

$$Adaptability\ Risk = \sum[Cost \times Difficulty]_{no\ propagation} \tag{3}$$

$$Extensibility\ Risk = \sum[Cost \times Difficulty]_{with\ propagation} \tag{4}$$

$$Complexity\ Risk = \left(\frac{new\ no.\ of\ part}{existing\ no.\ of\ part}\right) + \left(\frac{new\ no.\ of\ interrelationship}{existing\ no.\ of\ interrelationship}\right) \tag{5}$$

The score of each metric is then normalized by dividing it with maximum possible penalty score as indicated by Eqn. 6. In the calculation of overall evolvability risk, each metric can be assigned with a different weightage as in Eqn. 7 to mark the importance level of each assessment metric according to the preference of the designer. The baseline product design candidate with the lowest evolvability risk score can be taken as the best.

$$Normalized\ Risk,\ x_i = \frac{Total\ Risk\ Score}{Maximum\ Possible\ Risk\ Score} \tag{6}$$

$$Overall\ Evolvability\ Risk,\ f = \sum w_i x_i \tag{7}$$

To better demonstrate the baseline product design evaluation process, along with the consideration of change impact uncertainties, it is shown through a redesign case study of an aircraft system.

3.1. *Aircraft redesign case study*

This redesign case study is presented to demonstrate the capability of the baseline assessment method in selecting the best baseline product design with regards to redesign tasks at hand. In this case, three candidate aircraft are being considered: Airbus A320, Lockheed L-1011 and Boeing 727. The redesign changes are driven by the need to implement the electrical-mechanical actuator (EMA) for the primary roll control mechanism. All of the existing baseline candidate aircraft systems are using hydraulically-operated actuators instead of EMA. The implementation of EMA will help to reduce the total aircraft weight. A MATLAB program has been developed to calculate the overall evolvability risk and it runs with the user inputs such as on the aircraft system architecture, and change cost and difficulty ratings for each of the considered subsystems.

For this case study, three aircraft design performance parameters are taken as design requirements: flight range, maximum passengers capacity and takeoff gross weight. Table 4 presents the assessment scores for generality and scalability metrics for all of the considered baseline aircraft candidates. For instance, the generality score for Airbus A320 aircraft is assigned as 100 with respect to flight range requirement because the current flight range of this aircraft is just 3000 nm compared to the required 3900 nm. It is impossible for this required performance improvement can be achieved by the existing design without requiring any system modification, hence the penalty score of 100 is assigned. On the other hand, for the same design requirement, the scalability score is assigned as 0 since it is possible to achieve improved flight range by scaling down the design. Note that each of the design requirements is assessed independently and should not be influencing the scores for other requirements.

For adaptability and extensibility assessment, all initiating change components of the candidates need to be first identified. In order to do this, the architectural design of the candidates has to be well-defined. In this case implementation of EMA within the candidate's primary roll control mechanism, other components that changed due to earlier changed components also need to be identified. In other words, propagation of the changes of the components has to be identified. These connections between the components can be mapped into the design structure matrix (DSM). An example for Airbus A320 aircraft as presented in Figure 1, where the symbol "X" in the DSM indicates the existence of a change relationship between components of both respective row and column. This indicates that changing the component of the column will subsequently induce a modification on the component in the row. For example, changing the left electro-hydraulic servo jacks will cause some modification on elevator and aileron computers component of Airbus A320.

Table 4. Assignment of generality (G) and scalability (S) scores

Requirement	Target	Airbus A320		Lockheed L-1011		Boeing B727-100	
		G	S	G	S	G	S
Flight range (nm)	> 3,900	100	0	0	0	100	0
Maximum passenger capacity	> 234	100	0	0	0	100	0
Takeoff gross weight (lb)	< 255,000	0	0	100	0	0	0
Normalized Score		2/3	0	1/3	0	2/3	0

		1	2	3	4	5
Captain Sidestick Controller	1					
First Officer Sidestick Controller	2					
Elevator and Aileron Computers	3	X	X			
Left Electro-Hydraulic Servojacks	4			X		
Right Electro-Hydraulic Servojacks	5			X		

Figure 1. DSM for primary roll control of Airbus A320

Based on system design architecture presented by the DSM, possible initiating change components in the implementation of EMA are identified and their change difficulty and cost ratings are assigned. Table 5 tabulates the rating assignment for the Airbus A320 system based on the rating scales in the previous Table 2 and Table 3. In a similar fashion, the change difficulty and cost ratings for the other two baseline aircraft candidates are also individually tabulated in Table 6 and Table 7.

Table 5. Change difficulty (D) and cost (C) ratings for initiating change components of A320

Initiating Change	Change Remark	Adaptability Assessment		Extensibility Assessment	
		D	C	D	C
Left Electro-Hydraulic Servojacks	Changed to EMA	4	6	4	4
Right Electro-Hydraulic Servojacks		4	6	4	4

Table 6. Change difficulty (D) and cost (C) ratings for initiating change components of B727

Initiating Change	Change Remark	Adaptability Assessment		Extensibility Assessment	
		D	C	D	C
Left Inboard Aileron Hydraulic Actuators	Changed to EMA	10	10	6	4
Left Outboard Aileron Hydraulic Actuators		10	10	6	4
Right Inboard Aileron Hydraulic Actuators		10	10	6	4
Right Outboard Aileron Hydraulic Actuators		10	10	6	4
Aileron Power Control Unit	Changed to electrical	10	10	4	1

Table 7. Change difficulty (D) and cost (C) ratings for initiating change components of L-1011

Initiating Change	Change Remark	Adaptability Assessment		Extensibility Assessment	
		D	C	D	C
Master Aileron Hydraulic Actuators	Changed to EMA	10	10	6	4
Left Outboard Aileron Hydraulic Actuators		10	10	6	4
Right Inboard Aileron Hydraulic Actuators		10	10	6	4
Right Outboard Aileron Hydraulic Actuators		10	10	6	4
Master Outboard Aileron Servo	Changed to Electrical	10	10	4	1
Left Outboard Aileron Servo		10	10	4	1
Right Outboard Aileron Servo		10	10	4	1
Right Inboard Aileron Servo		10	10	4	1

To include uncertainty consideration into the change impact assessment, the assigned rating value is set as the mode value whereas 0 and 10 are taken as the lower and upper limits, respectively. With these known values, the best representative distribution for both change difficulty and cost ratings is a triangular distribution. Figure 2 depicts the triangular distribution for change difficulty and cost ratings with mode values of 4 and 6, respectively. Meanwhile, Figure 3 illustrates the resultant distribution of the product of the change difficulty and cost ratings. The adaptability and extensibility risk evaluation for all baseline aircraft candidates are calculated using the developed MATLAB computer program with the uncertainty consideration and the results are depicted in Figure 4.

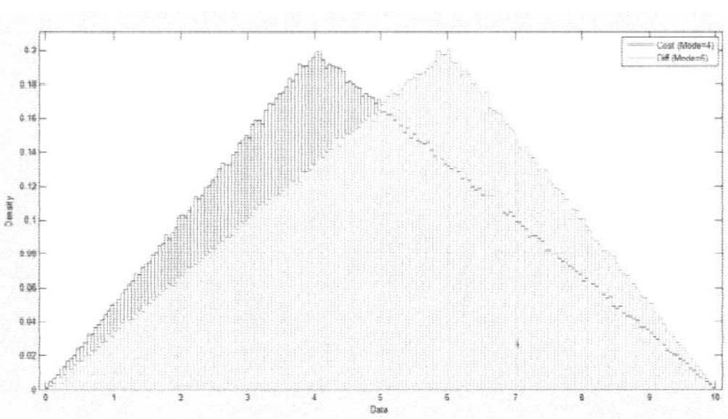

Figure 2. Example of a triangle distribution of cost rating with mode = 4 and difficulty rating with mode = 6

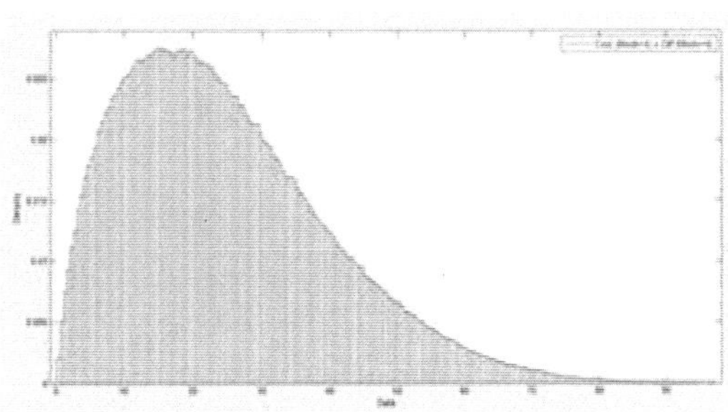

Figure 3. Example resultant distribution of product of cost and difficulty rating distributions in Figure 2

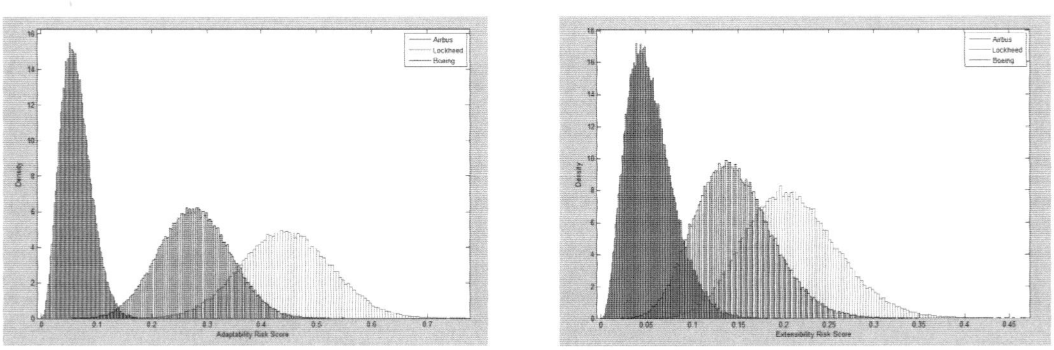

(a) (b)

Figure 4. Resultant output distribution for (a) adaptability and (b) extensibility risk scores

From Figure 4, it can be observed that there is an overlapping of risk scores between the considered baseline candidates, particularly for the extensibility metric, which is a sign that the aircraft might all have similar redesign risk level. Nonetheless, it is clear that Airbus A320 has the highest probability as the best baseline choice for this particular redesign task. Furthermore, Figure 5 presents the overall evolvability risk distribution for each aircraft that is computed using Eqn. 7 with an equal weightage for each of the five baseline assessment metrics (i.e. $w = 0.2$) at 1,000,000 random simulation runs. It can be observed that Airbus A320 is definitely the most potentially less risky for the redesign task as its overall evolvability risk score values are concentrated within the lower side of the risk distribution and hardly overlaps with the distribution of the other two. This makes it a robust choice as the baseline design for the redesign process.

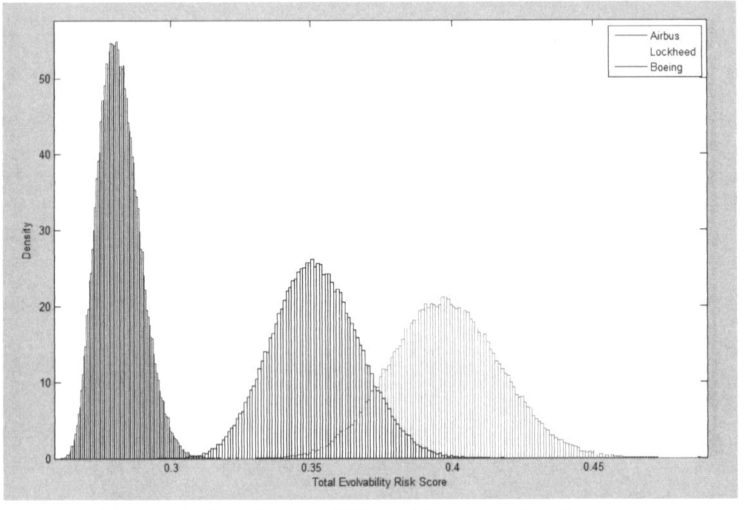

Figure 5. Total evolvability risk score distributions

3.2. Discussion

It can be seen from the results that Airbus A320 has the most potential for the baseline aircraft design to be used in the redesign case study, whereby its system design architecture is flexible and suitable in the EMA implementation process. The overall evolvability risk score distribution for the Airbus A320 is also not widely spread out and also highly concentrated with values in the lower risk region. These are good indicators that the aircraft redesign risk is of less uncertainties and highly probable to be of a low value. On contrary, distributions for evolvability risk score for both Lockheed L-1011 and Boeing B727 are rather spread out over the high values, indicating a high variability that might be harder to be comfortably predicted.

4. Conclusion

In this paper, the proposed robust baseline design selection methodology has been demonstrated using a simple aircraft redesign case study. Unlike static calculation for adaptability and extensibility metrics that have been done before, the impact of uncertainties of the change difficulty and cost ratings for the redesign task has been taken into account in this method. The resultant overall evolvability risk score distribution can provide more insights to the designer on the robustness of the baseline candidate to be used for the redesign development. This enables a better selection of the baseline design, hence leads to a more efficient product design and development process.

Acknowledgement

The authors acknowledge the financial support for this study from the Ministry of Higher Education, Malaysia through Exploratory Research Grant Scheme – ERGS/1/11/TK/UPM/02/12

IOP Conf. Series: Materials Science and Engineering **405** (2018) 012015 doi:10.1088/1757-899X/405/1/012015

References

[1] Eckert C M, Stacey M, Earl C. References to past designs. Studying Designers '05, Key Centre of Design Computing and Cognition, 2005; 3–21
[2] Regli W, Cicirello V A. Managing digital libraries for computer-aided design. Computer-Aided Design 2000; 32(2): 119-132
[3] Sawhney M S. Leveraged high-variety strategies: From portfolio thinking to platform thinking. Journal of the Academy of Marketing Science 1998; 26(1): 54-61
[4] Romli F I, Wiriadidjaja S, Mohd Rafie A S. A preliminary study of baseline design architecture effects on aircraft redesign risks. Applied Mechanics and Materials 2012; 225(2012): 287-292
[5] Karthik K, Reddy K. Engineering changes in product design - a review. IOP Conference Series: Materials Science and Engineering 2016; 149: 012001
[6] Clarkson P J, Simons C, Eckert C. Predicting change propagation in complex design. Journal of Mechanical Design 2004; 126(5): 788-797
[7] Eger T, Eckert C M, Clarkson P J. Engineering change analysis during ongoing product development. International Conference on Engineering Design, 2007
[8] Romli F. Baseline selection in redesign process. Frontiers in Aerospace Engineering 2013; 2(2): 116-128
[9] Romli F I. A strategic planning methodology for aircraft redesign. PhD Thesis, Georgia Institute of Technology, 2009
[10] Christian J A. A quantitative approach to assessing system evolvability, NASA Johnson Space Center, 2004
[11] Willems B, Duflou J, Dewulf W, Seliger G, Basdere B. MAAP: A method to assess the adaptability of products. 3rd Intenational Conference on Design and Manufacture for Sustainable Development, 2004
[12] Romli F I. Aircraft redesign through strategic planning of engineering changes. International Review of Aerospace Engineering 2010; 3(6): 336-352
[13] Romli F, Hou C K, Junxian C. Subsystems change ranking methodology (SCRaM) for complex product redesign process. Advanced Materials Research 2011; 308-310(2011): 167-173
[14] Romli F I, Hou C K, Junxian C. CROPS - A decision-making aid tool for product redesign. Applied Mechanics and Materials 2012; 110-116(2012): 5136-5140
[15] Romli F, Harmin M Y. Use of Monte Carlo method to estimate subsystem redesign risk for complex products: aircraft redesign case study. Aircraft Engineering and Aerospace Technology 2015; 87(6): 563-570
[16] Coppendale J. Manage risk in product and process development and avoid unpleasant surprises. Engineering Management Journal 1995; 5(1): 35-38
[17] Clarkson J, Simons C, Eckert C. Change prediction for product redesign. International Conference on Engineering Design, 2001
[18] O'Neill J, Thakur N, Duus A. Technical risk assessment: a practitioner's guide. Defence Science and Technology Organisation Edinburgh (Australia), 2007

Institute of Physics
Dirac House, Temple Back
Bristol BS1 6BE UK

ISSN: 1757-8981
ISBN 978-1-5108-7333-9

CURRAN ASSOCIATES INC.
proceedings.com

5th International Workshop & Summer School on Plasma Physics 2012

Journal of Physics: Conference Series Volume 715

Kiten, Bulgaria
25 – 30 June 2012

Institute of Physics